Practices in Social Ecological Research

Andrea Rawluk · Ruth Beilin ·
Helena Bender · Rebecca Ford

Practices in Social Ecological Research

Interdisciplinary collaboration in 'adaptive doing'

Andrea Rawluk
School of Ecosystem and Forest
Sciences
University of Melbourne
Parkville, VIC, Australia

Ruth Beilin
School of Ecosystem and Forest
Sciences
University of Melbourne
Parkville, VIC, Australia

Helena Bender
School of Ecosystem and Forest
Sciences
University of Melbourne
Parkville, VIC, Australia

Rebecca Ford
School of Ecosystem and Forest
Sciences
University of Melbourne
Parkville, VIC, Australia

ISBN 978-3-030-31188-9   ISBN 978-3-030-31189-6   (eBook)
https://doi.org/10.1007/978-3-030-31189-6

This Palgrave Pivot imprint is published by the registered company Springer Nature
Switzerland AG
The registered company address is: Gewerbestrasse 11, 6330 Cham, Switzerland

We wish to dedicate this book to all of those seeking to create change in and for social ecological systems and address social ecological justice through conscience, reflection, and reciprocity. We pay tribute to the next generation of activists, such as in 350.org and Climate Strikes, who will engage with the wicked challenges that social ecological systems present. From young to old, small actions to great activism, from local to global, your work matters and we look forward to being inspired by you.

This book is released in the same year as Budj Bim, an ancient aquaculture landscape in south-west Victoria, has been listed on the UNESCO World Heritage List. We dedicate this book to those who spent many years as activists working towards this outcome. This acknowledgement is just the beginning of rewriting a colonial narrative that has dominated for too long.

ACKNOWLEDGEMENTS

We thank the Gunditjmara people, the Traditional Owners of the land upon which we were honoured to meet and write while preparing this book. We pay respect to the elders of the community and extend our recognition to their descendants. We are grateful to the local people who generously shared their time and stories with us about *Tarerer*/Kelly Swamp. Any errors in the book are our own. Our achievements were enhanced by great catering and a fabulous view, for which we are very thankful. We are extremely grateful to Dr. Sara Maroske, Ana Lambert, and an anonymous reviewer for their detailed and invaluable feedback on earlier versions of the manuscript. We are grateful for the patience, open-heartedness, and contrariness of our co-authors and their ongoing willingness to enter the confronting space of collaboration.

ACKNOWLEDGMENTS

CONTENTS

List of Figures

Finding Ourselves in the Messy Entanglement of Complexity: An Introduction to the Challenges and Opportunities in Social Ecological Systems

CONTEXT IN WHICH WE WRITE THE BOOK

Social ecological systems (SES) and sustainability research offer exciting approaches to engage with the complex issues of our time, ranging from a single community protecting a beloved local area from development, to management of a state conservation area, to the impacts of the Anthropocene that ricochet across global and local scales. Recognition of interdependent relationships between humans and the environment has been essential to the advances made in SES research, as has the acknowledgement that SES are non-linear and dynamic. A key challenge is finding integrated approaches to SES that combine the knowledge and practices from the many disciplines that contribute to this space.

Different scholarships have made advances in helping explicate the range of ways we might think about and engage with the challenge of integrating knowledge. However, the complexity remains, and this may reflect difficulties in achieving interdisciplinarity. Some scholars, including ourselves, argue that this integration of what and how we know SES is incomplete (Cumming 2014; Herrero-Jáuregui et al. 2018). Disciplinary and philosophical differences, even if unconscious, are often irreconcilable (Phoenix et al. 2013). Tackling integration by focusing on more abstract philosophical and disciplinary differences can

A. Rawluk et al., *Practices in Social Ecological Research*,
https://doi.org/10.1007/978-3-030-31189-6_1

present an obstacle for researchers, practitioners, and students interested in SES. Instead, we suggest that starting with practice, starting with what researchers, practitioners, and people do in their everyday lives or disciplines, and using narratives as accessible stories, can act as a doorway to reflect on what is known and prioritised, along with engaging with more abstract differences. This book supports taking action by outlining a practice-focused way to navigate the messiness of social ecological challenges, and serves as a vehicle for empowerment, vision, and action at a time when there is an increasing number of complex issues that threaten human survival and demand approaches that can facilitate sustainability.

Why This Book

Interdisciplinary collaboration, or synthesis, is at the core of SES research and management, sustainability science, and many other areas. However, integration is a messy business, especially in SES because the synthesis process needs to occur at multiple scales: how learnings integrate with policy, how the frameworks, tools, and practices of social and ecological disciplines can be brought together, as well as how our individual practices as researchers and practitioners need to respond to changing contexts and the integration of new learnings. There is a wide array of SES frameworks that are effective within a particular discipline, but these often have limitations in their ability to link and integrate SES together (Binder et al. 2013; Cumming 2014), and there are differences of opinion about how to integrate social ecological knowledge. Herrero-Jáuregui et al. (2018) called for a well-documented framework to build bridges between the disciplines connected to social and ecological ways of knowing. Recent reviews of SES research suggest that synthesis remains a challenge, of bringing together different knowledges (Cumming 2014), bringing together different practices (Herrero-Jáuregui et al. 2018; Perz 2019), as well as exploring the relationship between epistemology (how we know), ontology (how we view reality) and axiology (our values) (Binder et al. 2013; Cumming 2014; Collard et al. 2018).

This book seeks to actively, and with humility, engage with the challenge of integration within the social ecological systems research and management space. We focus our energies here because we are concerned by complex, social ecological challenges. Through understanding and responding to SES, we wish to contribute to efforts to facilitate sustainability. In research on how to express and to integrate multiple human and non-human aspects in SES, the starting point is often grappling with different knowledges (Phoenix et al. 2013), which are inherently irreconcilable and can make processes initially seem futile. As we indicated above, we offer a reorientation of interdisciplinary integration by turning to practice. However, there is no pre-existing map for navigating this kind of interdisciplinary practice in SES thinking that is accessible to practitioners and disciplinarians alike. We needed a map ourselves, so we developed this book.

'Alter your perspective by a few degrees, and the view is different' Bruce Pascoe (2014, p. 36) states in *Dark Emu*, on seeing Indigenous history and culture in Australia disentangled from colonial racism. We turn to practice in this book as a starting point for developing awareness of and then shifting our view, our conscious engagement in interdisciplinarity, and our participation in the struggle for social and environmental justice. In response, we offer the practice-oriented process of 'adaptive doing' in which people are asked to do differently, see differently, and open space for unexpected outcomes to emerge.

The aims of this book are:

- to outline and demonstrate 'adaptive doing', a practice-oriented process for integrating research in SES, that is transparent, inclusive, and engaged;
- to demonstrate three reframing tools from the social sciences—the 4 Is, assemblage, and the eternally unfolding present—that assist SES researchers and practitioners to participate in 'adaptive doing'; and
- to overcome disciplinary silos by creating a platform that we call the 'agora', which creates a space where SES researchers and practitioners can participate in 'adaptive doing' to learn and improve SES practices and outcomes.

In response to these aims, we offer a practice-focused approach that draws on a breadth of scholarship across SES thinking, interdisciplinarity, social learning, and critical reflection. By practice, we mean any kind of ongoing, often everyday activity that involves a combination of knowledge and context as constituent parts (Cook and Wagenaar 2012). Given our focus on SES research, we are particularly interested in practices that are contributing to, elucidating, or mitigating complex issues related to sustainability.

Approach Taken in the Book

We take an interdisciplinary approach in this book. We see interdisciplinarity as a process for co-creating shared understanding of a phenomenon or system that shapes and is shaped by those involved. We, the authors, come from different disciplinary backgrounds, although we all started our disciplinary training within the natural sciences. We seek to be transparent, owning the biases that we bring to writing and engaging in social ecological research. We come from and/or work across different ontological positions, including: post-positivism, perceiving that there is an imperfectly knowable single real world; constructivism, wherein there are multiple understandings of the world, which are known through each person's experience and are built over time; and critical theory, which sees multiple understandings of the world, and acts in the world to illuminate and create change. We seek to bring a just and ethical approach in the processes and examples we offer in this book.

Who Is the Book for?

We write this book for other researchers and practitioners who work or are interested in a systems-thinking approach for engaging with SES issues. We see systems as a network of relationships that form an integrated whole, that are nested within other systems, and contain subsystems (e.g., Berkes et al. 2003). We recognise that not all researchers in areas of sustainability or environmental issues engage with systems thinking, but we welcome such researchers to explore this approach along with critical reflection. We write this book for people who are interested in alternative ways of thinking about sustainability and opening new questions and directions for practice.

Structure of the Book and the Function of Each Chapter

In this first chapter, we outline the impetus for the book and how and to whom we think it can be useful. In the second chapter, we introduce a wetland case study that at first appears unremarkable but is itself full of drama. Our case is situated in south-west Victoria, Australia and we return to it in the following chapters. In Chapter 3, we utilise our case to highlight examples of the achievements and challenges currently faced in SES research and management. Chapter 4 introduces three key elements: 'adaptive doing', which is a practice-oriented process; a platform in which to participate in adaptive doing, the 'agora'; and three reframing tools—the 4 Is, assemblage, and the eternally unfolding present, that offer different perspectives. The 'adaptive doing' process works to develop awareness of changed positions and improve integration, which leads to changes in understanding and practices among local communities, by researchers and with practitioners, and offers new insights that can assist us to facilitate sustainability. The 'agora' provides a space and time in which a practice-oriented approach can occur, it assists in building mutual respect among participants, and to overcome path dependencies in an SES. The three reframing tools that come from the social sciences, assist integration by offering different perspectives. Finally, in Chapter 5 we share the outcomes of applying adaptive doing and the three reframing tools to the case. We conclude with reflections on the insights we gained as researchers and practitioners from being in the 'agora' and engaging with the adaptive doing process.

References

Berkes, F., J. Colding, and C. Folke (eds.). 2003. *Navigating Social-Ecological Systems: Building Resilience for Complexity and Change*. Cambridge, UK: Cambridge University Press.

Binder, C.R., J. Hinkel, P.W.G. Bots, and C. Pahl-Wostl. 2013. Comparison of Frameworks for Analyzing Social-Ecological Systems. *Ecology and Society* 18 (4): 26.

Collard, R.-C., L.M. Harris, N. Heynen, and L. Mehta. 2018. The Antinomies of Nature and Space. *Environment and Planning E: Nature and Space* 1 (1–2): 3–24.

Cook, S.D.M., and H. Wagenaar. 2012. Navigating the Eternally Unfolding Present: Toward an Epistemology of Practice. *The American Review of Public Administration* 42 (1): 3–38.

Cumming, G. 2014. Theoretical Frameworks for the Analysis of Social-Ecological Systems. In *Social-Ecological Systems in Transition*, ed. S. Sakai and C. Umetsu. Otsu, Japan: Springer.

Herrero-Jáuregui, C., C. Arnaiz-Schmitz, M. Fernanda Reyes, M. Telesnicki, I. Agramonte, M.H. Easdale, M. Fe Schmitz, M. Aguiar, A. Gómez-Sal, and C. Montes. 2018. What Do We Talk about When We Talk about Social-Ecological Systems? *A Literature Review Sustainability* 10 (2950): 1–14.

Pascoe, B. 2014. *Dark Emu: Aboriginal Australia and the Birth of Agriculture*. Broome, WA: Magabala Books Aboriginal Corporation.

Perz, S.G. 2019. Crossing Boundaries for Collaboration in Comparative Perspective: Key Insights, Lessons Learned, and Recommendations for Future Practice. In *Collaboration Across Boundaries for Social-Ecological Systems Science*, ed. S. Perz. London, UK: Palgrave Macmillan.

Phoenix, C., N.J. Osborne, C. Redshaw, R. Moran, W. Stahl-Timmins, M.H. Depledge, L.E. Fleming, and B.W. Wheeler. 2013. Paradigmatic Approaches to Studying Environment and Human Health: (Forgotten) Implications for Interdisciplinary Research. *Environmental Science and Policy* 25: 218–228.

CHAPTER 2

Engaging with a Social Ecological System: The Swamp

Throughout this book, we focus on a swamp in south-west Victoria to illustrate ways of engaging with a social ecological system (SES) through the practice-oriented approach of adaptive doing. We draw on examples from this locality to illustrate gaps in how we engage in practices in SES. In this chapter, we present the SES in a conventional way and summarise a collection of narratives that were gathered from people who live and work around or near this site. In doing so, we prepare the reader to interrogate the way we understand the system.

How boundaries are set can have profound implications for who and what is included and excluded from an SES. We discuss the setting of boundaries in more detail in Chapter 3. As a place to start, we base the boundary of this swamp on the geography of the landscape, such that the local system centres on a swamp named on government maps as Kelly Swamp that runs for several kilometres behind, and parallel to tall sand dunes that separate it from the beaches and the Great Southern Ocean. The eastern end of the Swamp is intersected by the freshwater Merri River on its way to Armstrong Bay west of the regional city of Warrnambool. On the western end of our system is Rutledge Cutting and Saltwater Swamp (Fig. 2.1). At Rutledge Cutting, the sea comes into the dune and estuary system. At the Merri River end, freshwater enters the Swamp under flood conditions and goes out to the sea, when the river mouth is not closed by sand. Kelly Swamp is of national significance for its important geomorphologic and ecological features (Belfast Glenelg Hopkins Catchment Management Authority 2008).

© The Author(s) 2020
A. Rawluk et al., *Practices in Social Ecological Research*,
https://doi.org/10.1007/978-3-030-31189-6_2

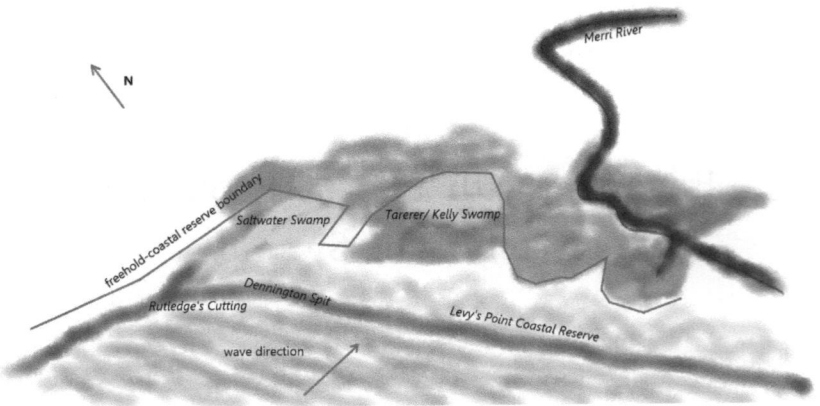

Fig. 2.1 Map of Kelly Swamp and the immediate surrounds (*Note* Visual depiction of *Tarerer*/Kelly Swamp. Although not to scale, two centimetres represents approximately 500 metres. The blue represents the ocean. The orange line depicts the beach with Dennington Spit. The yellow represents the sand dune formation of Levy's Point Coastal Reserve. The green represents the pastoral elements of the landscape. Within the pastoral elements are *Tarerer*/Kelly Swamp and Saltwater Swamp. There is a freehold-coastal reserve boundary that cuts through this space. The Merri River is represented in darker blue) (Color figure online)

THE BIOPHYSICAL SETTING

The geological formation of Kelly Swamp is part of the history of the ancient continent of Australia that separated from Gondwanaland about 180 million years ago. During the last Ice Age (which ended about 11,000 years ago), the sea in this region was about 120 m lower than it is today and the coast was some 50 km farther out on the continental shelf. This means the water flows associated with this land are a relatively recent phenomenon. It wasn't until some 6000 years ago, when the climate warmed and the ice caps melted, that a two-metre rise in sea level above today's level inundated the area where the Swamp is now located. The sea retreated again, as evidenced by fossils and a shell bed dated 5520 ± 70 years B.P. (Gill 1978, p. 42). The Dennington Spit—which is now part of the ongoing sand dune formation—began to build as the sea retreated, creating the Merri estuary on the north side of the Spit and a much lower version of the current sand dunes. The Merri

River had to flow around the tip of the Spit (to the east) but as the Spit grew the river was diverted in 1859, partially exiting through a canal in Warrnambool. A large basalt flow intersects the main water channel between Kelly Swamp and the more western Saltwater Swamp as water moves towards Rutledge's Cutting (Glenelg Hopkins Merri Estuary MP 2008). This structure can hold back freshwater on the east side, except in higher river flows when the structure is overtopped, affecting the salinity in the two Swamps (Fenton 2005, cited by Glenelg Hopkins Merri Estuary MP 2008).

Kelly Swamp lies within the Hopkins River basin system (EPA 2004). The total length of the Merri River is about 110 km with a catchment of approximately 19000 km^2 (EPA 2004). The flows in the river are historically highly variable, with highest flows between June and December, and low flows between January and May (EPA 2004). The Merri River carries both dissolved and particulate organic matter, as well as nutrients collected along the length of the river, into the estuary, which is vital for the growth of estuarine plants and animals (EPA 2011). The Swamp varies in width from 500–1000 m and covers an area of 1.46 km^2 (Glenelg Hopkins Merri Estuary MP 2008).

Gill (1978) estimates that the sand dunes along the foreshore between the beach and Kelly Swamp were smooth and stable for 2800 years prior to European occupation. They were covered in soil to about 0.5 m or more thickness as well as indigenous vegetation. The slopes of the dunes were about 10 degrees or less, such that their maximum height was 3 metres (Gill 1978). The coastal formation on the ocean side of Kelly Swamp (including the incorporation of Dennington Spit) is a stranded dune comprised of self-cementing calcarenite. The formation of this Spit effectively means that Kelly Swamp becomes a reservoir for floodwaters. Gill (1984) estimates the age of the Spit to be 13,000 years.

There are two distinct aspects of the Warrnambool coast (Gill 1984). The first is the swell wave regime, which is significant in providing waves that arrive parallel to the shore (see Fig. 2.2). Secondly, sand travels along the coast, lifted by the waves off the ocean floor. The sand at the mouth of the Merri River is not a delta formation, with sand brought to the sea by the river, but rather sand accumulation is caused by 'constructional waves', with sand brought to the mouth of the river from the sea (Gill 1984, p. 15). During the dry summer months when there is little water flowing in the river, the sand is deposited at both exits to the swamp, the Merri River Cutting and Rutledge's Cutting. The sand

Fig. 2.2 Aerial view of Kelly Swamp looking north-east over Kelly Swamp towards Dennington and Warrnambool (Photo credit: Chris Farrell Nature Photography, June 2017)

deposition can completely block Rutledge's Cutting (Glenelg Hopkins Merri Estuary MP 2008). After the winter rains start, the Merri River floods, and significant water is held back at the mouth by this closure, until the cutting is eventually opened again (see Fig. 2.3).

The Otway, Highland, and Murray groundwater systems underlie the region. The majority of the Merri Catchment sits atop the Yangery aquifer, which is unconfined and thus influenced by rainfall and atmospheric conditions (EPA 2004). It is estimated that it contains about 11,000 ML of water (EPA 2004). The area experiences approximately 800 ml of rain a year, with a Mediterranean climate, characterised by wet and cool winters and hot and dry summers. This region has experienced and continues to experience many biophysical processes that have powerfully transformed its material form, including climate change.

The area around Warrnambool has been described as covered by the following ecological vegetation classes prior to 1750: damp sands, herb-rich woodlands, coastal dune scrub, swamp scrub, aquatic herb lands, and estuarine wetlands (Glenelg Hopkins Merri Estuary MP 2008).

Fig. 2.3 Aerial view of Kelly Swamp looking west over Rutledge Cutting in flood (Photo credit: Chris Farrell Nature Photography, September 2017)

Reed beds of (*Phragmites australis*) and spiky club rush (*Schoenoplectus pungens*) provide habitat, breeding areas and feeding grounds for birds, reptiles, and some fish.

Kelly Swamp is important for conservation as there are 30 waterbird species that have been recorded there, with 15 of these species listed in the Japan-Australia Migratory Bird Agreement and the China-Australia Migratory Bird Agreement (GHCMA 2008). It contains significant winter habitat, such as glassworts (*Sarcocornia quinqueflora*), for critically endangered species such as the orange-bellied parrot (*Neophema chrysogaster*). The beach side of the dunes provides breeding grounds for the rare hooded plover (*Thinornis rubricollis*), along with other ground-nesting birds, and the area serves as an important refuge during drought (GHCMA 2008). It is also in the migration path of the black swan *Cygnus atratus* (Fig. 2.4). In addition to avifauna, there are a number of other species that utilise this area: the rakali or water rat (*Hydromys chrysogaster*), which inhabits banks along rivers, lakes, estuaries, and marine shores (McGregor 1995), and short-finned eels (*Anguilla australis*) are examples.

Fig. 2.4 Diverse bird life at *Tarerer*/Kelly Swamp (Photo credit: Ana Lambert, February 2018)

THE SOCIAL SETTING

Kelly Swamp and the surrounding areas have been inhabited by Indigenous clans for over 60,000 years (McNiven et al. 2015), which includes the last Ice Age. Dhauwurd wurrung, Djab wurrung, and Girai wurrung are language groups, belonging to the Gunditjmara or Eastern Marr people, who have had continuous occupation of this region (Clark 1995) (see Fig. 2.5). Kelly Swamp is known as *Tarerer* in local Gunditjmara language. *Tarerer* ('Kelly Swamp') was a major meeting place for hundreds of people from the Indigenous nations, which is described by Clark (1995) as a swamp between Tower Hill and Merri River. The Murrnong, a yam daisy staple cultivated by Indigenous people across the western plains, grows naturally around swamps (Doyle 2006). Similarly, Indigenous eel fishing and trapping is associated with estuary areas along this coast (McNiven et al. 2015) and the nearby Budj Bim eel trap complex has been recognised as a World Heritage site in 2019 by UNESCO. There are Indigenous people's middens present in the thin layer of soils that overlay the dunes (Gill 1954, 1987).

The date of the first European visit to the area is a matter of conjecture. Rumours of the Portuguese coming in the fifteenth century are

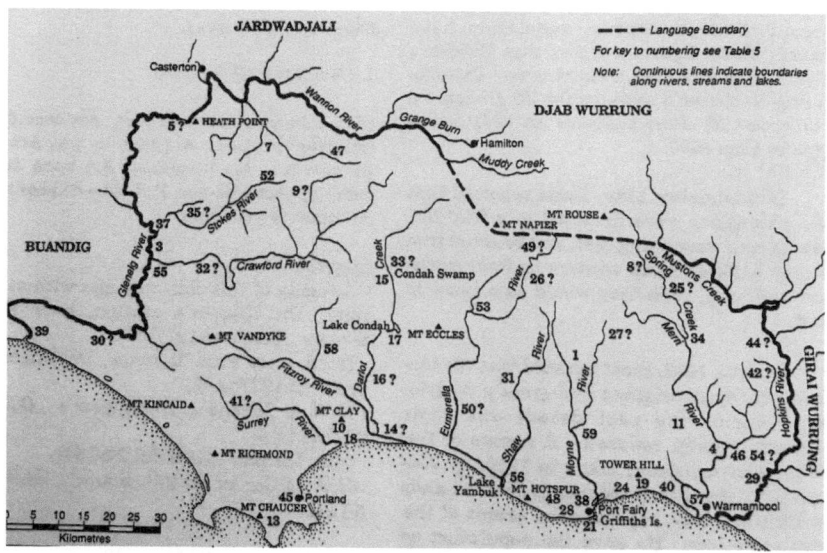

Fig. 2.5 Map of Dhauwurd wurrung (Gunditjmara) dialects and where they occur relative to the different rivers in south-west Victoria. On this map, *Tarerer*/Kelly Swamp is approximately located around the number 40 (from Clark 1990)

associated with the 'Mahogany Ship'. The ship is allegedly buried in the sands between the Merri River and Gorman's Lane at Rutledge Cutting. It was discovered by whalers, among others, in the mid-1800s (Powling 2003), but has eluded continuing efforts to locate it and so a great deal of mystery surrounds its origins and time of arrival.

The first colonisation of the area by Europeans was in 1827. Many came to the western districts of Victoria from Tasmania, claiming land illegally as squatters (Builth 2009). Some Gunditjmara clan peoples were massacred in 1887 at the Convincing Ground near Portland (Clark 1995). Settlers introduced European disease, and this compounded by changed diets and some tribal conflict were responsible for 70% of the Aboriginal population being killed, disappeared or removed between 1835 and 1845. Those that survived were relocated to missions at Lake Condah (Budj Bim), Framlingham (in south-west Victoria) and Coranderrk, north-west of Melbourne (Builth 2009).

Fig. 2.6 View across *Tarerer*/Kelly Swamp showing farming and sand dunes (Photo credit: Rebecca Ford, January 2019) (Color figure online)

Undoubtedly, the European arrivals noted the green landscapes of the floodplains. They interpreted these as a promising indication of available water and that this land might be good for European-style agricultural living (see Figs. 2.6 and 2.7). It is reported that almost immediately, the new settlers sought to create the perfect pasture for sheep, which began with a forensic clearing of native vegetation, such that approximately 95% of the native vegetation was removed (McGregor 1995). The soil and land condition in Hopkins water catchment, of which Kelly Swamp is part, was rated as marginal and poor in 2004 because of flow deviation, degraded riparian vegetation, introduced fauna, especially predators, loss of in-stream habitat, and persistent stock access (GHCMA 2008).

In the 1840s, William Rutledge headed a syndicate of northern Irishmen who purchased land between the Merri River and Saltwater Swamp at Killarney encompassing 5000 acres (Doyle 2006). This is the area of about 15 km², which contains Kelly Swamp and the Merri River

Fig. 2.7 View across *Tarerer*/Kelly Swamp showing the Merri River with cattle grazing on adjacent pasture in the swamp landscape (Photo credit: Rebecca Ford, January 2019) (Color figure online)

estuary (see Fig. 2.1). The Merri River estuary is linked to the Great Southern Ocean through the Merri Marine Sanctuary, which was established in Stingray Bay in 2002 (GHCMA 2008).

Swamplands were thought of as unproductive by the European settlers and from the 1870s there was a concerted effort to drain them across this region (Doyle 2006). The intention was to dry out the floodplain in order to make larger fields or pastures. Settlers also began extracting groundwater through pumping for their farming and grazing. Irrigation pumping from the Merri River commenced in 1945 (EPA 2004). Over time, groundwater extraction reached 17,000 ML (EPA 2004), far exceeding the estimated available volume of water in the Yangery aquifer underpinning the landscape. Water extraction coupled with the loss of deep-rooted vegetation had direct impacts on the groundwater flows into the region's rivers and wetlands (Tweed et al. 2007), although the intermingling of surface and ground water complicates visual assessments of water in the landscape. In contemporary times, local farmers say that the water is just a shovel depth below the surface along the coastal estuary complexes, and this includes the floodplains along the Merri River and Kelly Swamp.

In 1859 the Merri Cutting (a diversion canal for the Merri River that was intended to ensure that land did not flood and that water could easily escape to the sea) was built, based on the false premise that the sand that blocked the river mouth in late spring-early summer came from the hinterland as a result of river flooding. Mechanical opening of the river mouth at Rutledge Cutting has been justified for economic and social values (GHCMA 2008), despite recognition that there can be negative ecological impacts including fish deaths (EPA 2011).

The Glenelg Hopkins Catchment Management Authority (GHCMA 2008) is responsible for monitoring and managing the water, vegetation, and land including Kelly Swamp. They require farmers to fence and restrict cattle access to the Merri River, and to assist with plantings to encourage landward migration of vegetation with sea-level rise, as well as reduce riparian erosion. The GHCMA also set the protocols for the opening of Rutledge's Cutting, although this is implemented by Parks Victoria—a government entity. There are disputes between these government agencies and the farmers about when the water moving through Rutledge's Cutting should be allowed to flow, and whether the cutting should be reopened manually (GHCMA 2008).

Gill (1984) notes that the current great height of the dunes and the steepness of the slope is due to marram grass (*Ammophila* spp.) trapping sand. Planting of marram grass, a Western European species, began in 1890 to stop sand blowing into farmers' pasture on the floodplain (Heathcote and Maroske 1996). The marram grass also created irregular surfaces in the dunes into which early farmers put their cows for shade and feed in hot weather. Their hoofs caused erosion and their paths through the dunes created significant tracking lines. Locals were concerned that the dunes were responsible for sand blowing on to the cleared farmland and understood that marram grass would help the dunes stabilise. Another consequence of the introduction of marram grass by local farmers has been an increase in the height of the sand dunes to between 4–8 m, which is one-third to two times as high as it was before the Europeans arrived in the 1800s (Gill 1984).

At the time of writing, part of the freshwater Kelly Swamp that floods in winter and dries out in summer was primarily used for grazing beef or dairy cattle. For local farmers, this provides a source of feed in summer after pastures on higher ground dry out. Dairy farming, traditionally an important European land use in the area, has been in decline, while rural residential use of land is increasing, associated with increases in the population of the nearby town of Warrnambool. Use of the beach and river landscape for recreation has increased since 2012, when a rail trail was established on a decommissioned railway between Warrnambool and Port Fairy. The rail trail follows the Merri estuary before crossing Kelly Swamp and heading to the old highway to Port Fairy.

The area of Kelly Swamp expands and contracts according to the amount of water that seeps in from the land and falls from the skies and is clearly an ecosystem that is shaped by water and by people. The name of this area has changed over time, being known by the Gunditjmara as *Tarerer* and Kelly Swamp by the settlers. To honour this history, we use both names [*Tarerer*/Kelly Swamp] from here. The plants and animals that are found here are adapted to wet and dry seasons. The dry, from December through to April or May, leaves little and mostly no water on the surface in the years since our observations began (in 2015). The migratory (magpie geese, Latham snipe) and domestic birds (black swans and ducks most noticeably) disappear during the dry.

In short, the life of the Swamp reflects the dynamic realities of SES—non-linear, complex, presenting emergent possibilities and the uncertainties of climates, markets, time, and space. However, European occupants attempted to command and control the water, the use of the Swamp and its connections to the larger landscape. While it is not the sole purpose of this book to understand how the Swamp functions as a system, we will highlight the relationships between different elements and actors in this SES, as well as interactions between humans and the environment.

Example Narratives About *Tarerer*/Kelly Swamp

To extend our understanding of *Tarerer*/Kelly Swamp, we draw on a narrative approach, which is a common tool in the social sciences. The narratives we present here are based on conversations we had with people who live locally. We asked our participants to tell us three things about the Swamp: what they knew about the Swamp, what they envisioned as the likely future of the Swamp and the region in 100 years, and their desired future for the Swamp.

To protect the privacy of the local people that we talked with, while presenting readers the insights that a narrative approach offers, we provide summaries of our participants' positions, world views, and key observations about the Swamp below. In Chapter 5, we include a few excerpts from the larger discussion that our narrators shared, where relevant. Our goal in providing the summaries and stories is to demonstrate how narratives create the background for the 'adaptive doing' we discuss later in the book.

Scientists

Among our participants are three local scientists (an ecologist, earth scientist, and natural resource manager). They share some common expectations of the Swamp landscape that allow them to position themselves to it in different ways. Two of them describe an objective stance in which nature exists and can be observed and in which people can participate by passing through it (observing nature) or by actively monitoring in the area. All three of the scientists have been actively involved in some forms of ecological monitoring in this area.

> ...we monitor the salt content in the estuary so that the bream can satisfactorily lay their eggs...this is a naturally occurring cycle that is an indicator of the river and estuary health...

> ...we monitor the birds in this coastal region, there are many endangered and nearly extinct ...perhaps functionally extinct species already ... that rely on the wetlands and reeds for food and shelter...

> ...we study the plant vegetation communities...

From these, they construct mutually held understandings of the place so that their narratives are not dissimilar about the richness of species and the importance of nature to these places.

Their narratives also mention the importance of recreational use in preserving the habitats and species they study and the national and international conservation and habitat overlays across these wetlands. They consider climate change to be just one threat to the management of the area; with immediate concerns coming from the excision of public land along the beach allowing the horse racing industry to train up to 100 plus horses a day on the beach and dunes. Two scientists noted that climate change would alter the flow of water through the catchment affecting the migration of birds and their use of this habitat. One considers humans to be of little overall consequence to the future of the site.

For each of these individuals, science has given them a framework to see the landscape as replete with non-human species. They are all actively engaged in ongoing research as a lifework, and they all volunteer in various environmental causes that are subject to funding cycles and administrative constraints. They carefully make no criticism of their farming colleagues, rather proposing that the entrenched ways that each has of ordering their worlds breed mutual exclusivity.

Farmers

There are three farming families who shared their narratives with us. All of them begin their stories from the time of European settlement in the 1850s, calling upon socio-economic and legal customs around property rights. They all reflect on their family's historical acquisition of land in this area. All of them acquired their current property from family or in-laws within this current generation. They continue to farm land that was cleared for potatoes by their European antecedents. These landscapes became dairy country and now as this generation ages, it is gradually moving from dairy to grazing cattle and horse agistment. They sell their products nationally and internationally, relying on local agricultural production associations like the United Dairy Farmers, as well as trader and service companies like Landmark, which offer information on these markets.

All of these farmers consider production to occur in the pastures and paddocks alongside the river, its estuary, and swamp. In taking management actions, the farmers talked about acting collectively, like cutting a drain in the Swamp. The Swamp also serves as a kind of waterhole for the farm and wild animals as well as an aesthetic landscape when there is water. All the farmers describe the productive value of Kelly Swamp to their properties. It is a reason that these blocks of land are so closely held, as they are a significant advantage in the dry landscape of summer. One of the farms is biodynamic and the others use mainstream farming practices. They all have stories of watching the birds that suddenly appear, including old stories of hunting and eating magpie geese, and of the damage the birds can do to pasture and crops. They also note the ever-present relationship between the cattle and ibis (*Threskiornis moluccus*). The farmers locate themselves as benign observers of the non-human world, but there are clear demarcation lines between conservation (which is 'nice') compared to the importance of being productive, in all their stories. Some of the farmers mentioned that salinity was increasing in the estuary and swamp habitats closer to the shore. In one family, we note that one of the farming duo plants significant indigenous tree and shrub corridors across the floodplains.

One of the farmers tells us how their cattle used to roam the dunes (until the mid-90s when State conservation managers evicted them 'overnight') and the dangers to cows there because of snakes in the dunes. The swamp and dunes as habitat for snakes is a common association among the farmers. Each of them has memories of snakes killing horses (used in past years for rounding up cows), and snakes travelling up to the farmhouses from Kelly Swamp during the dry times. Therefore, for them, the swamp's seasonal habits are cautionary and they try to regulate the flows with drains.

Artists

There are two local artists who brought music, art, and stories to share with us. One of these has spent their life in this landscape. Growing up here means that they went to school with Gunditjmara children, many of whom remain friends with this artist in adulthood. This artist explained how Indigenous representation is divided between clan groups. The musician reflects a deep passion for the landscape and for the importance of reconciliation with Traditional Owners so that social and environmental justice can ensue. This artist has used music to champion local festivals that celebrate the life of the Traditional Owners as the people who live and belong here.

The second artist uses art to bring people together who are facing disruptions and shocks associated with land and water management along the coast. Undertaking this work has allowed this artist to overcome some of the barriers to changing land management practices such as restoration of creek and river tributaries and loss of property associated with flooding. This artist has also undertaken a creative visualisation of some of the coastal region that reimagines it as a coastal she-oak—the local tree—open woodland.

Through these performances, these artists help us to conceptualise past landscapes, the migratory bird stories, the beach, how climate change is altering these memorable and loved aspects, and to give form to a timeless experience of human and non-human coexistence that is radically different to the other narratives we hear.

The Gunditjmara

The voices of the Traditional Owners of *Tarerer* and its flood plains and dunes are missing from our first-hand narratives. We acknowledge their absence. We did arrange a meeting but, in the end, it did not occur, and we have relied on the historical record and the contemporary narratives of the locals to keep them present in this book.

In this chapter, we provided a conventional SES and narrative-based description of *Tarerer*/Kelly Swamp. We draw on this description in subsequent chapters to examine the assumptions and ideas common to SES thinking and to demonstrate how an 'adaptive doing' approach, based in practice, creates the pathways for alternative ways of being and working within SES. In Chapter 3, we review the SES literature and build a case for a focus on practice.

REFERENCES

Builth, H. 2009. Intangible Heritage of Indigenous Australians: A Victorian Example. *Historic Environment* 22 (3): 24–31.

Clark, I.D. 1990. *Aboriginal Languages and Clans: An Historical Atlas of Western and Central Victoria 1800–1900*. Melbourne, Australia: Monash University.

Clark, I.D. 1995. *Scars on the Landscape: A Register of Massacre Sites in Western Victoria 1803–1859*. Canberra: Aboriginal Studies Press.

Doyle, H. 2006. *Moyne Heritage Study Volume 2: Environmental History*. Prepared for Moyne Shire Council in association with Context Pty Ltd.

Environment Protection Authority (EPA). 2004. *Environmental Audit: Merri River Estuary Findings and Recommendations*. Melbourne: Environment Protection Authority.

Environment Protection Authority (EPA). 2011. *How Will Climate Change Affect Victorian Estuaries?* Melbourne, VIC: Environment Protection Authority.

Fenton, C. 2005. *Hydrodynamics and Nutrient Status of the Fitzroy River Estuary—Progress Report 5.* Warrnambool: Deakin University.

Gill, E.D. 1954. Aboriginal Kitchen Middens and Marine Shell Beds. *Mankind* 4 (6): 249–254.

Gill, E.D. 1978. *Quantification of Coastal Processes as a Basis for Coastal Management and Engineering.* CSIRO.

Gill, E.D. 1984. *Coastal Processes and the Sanding of Warrnambool Harbour. Definition of Coastal Processes, Quantification of Sand Erosion/Deposition, and the Reason for the Sanding Up of the Harbour at Warrnambool, S.W. Victoria, Australia* (editor). Warrnambool Institute Press.

Glenelg Hopkins Catchment Management Authority (GHCMA). 2008. *Merri Estuary Management Plan.* Hamilton, VIC.

Heathcote, J., and S. Maroske. 1996. Drifting Sand and Marram Grass. *The Victorian Naturalist* 113 (1): 13.

McGregor, J. 1995. *The Merri River—An Environmental Audit.* Warrnambool: Deakin University.

McNiven, I.J., J. Crouch, T. Richards, K. Sniderman, N. Dolby, and Gunditj Mirring Traditional Owners Aboriginal Corporation. 2015. Phased Redevelopment of an Ancient Gunditjmara Fish Trap over the Past 800 Years: Muldoons Trap Complex, Lake Condah, Southwestern Victoria. *Australian Archaeology* 81: 44–58.

Powling, J.W. 2003. *The Mahogany Ship: A Survey of the Evidence.* Warrnambool: Osburne Group.

Tweed, S.O., M. Leblanc, J.A. Webb, and M.W. Lubczynski. 2007. Remote Sensing and GIS for Mapping Groundwater Recharge and Discharge Areas in Salinity Prone Catchments, South-Eastern Australia. *Hydrogeology Journal* 15: 75–96.

CHAPTER 3

A Critical Reflection on Social Ecological Research and Turning to Practice

In this chapter, we build the case for focusing on practice as an entry-point to improving interdisciplinarity and struggles over reality, knowledge, and power in social ecological research. First, we examine the use of language as a starting point for arguing for intentional engagement in a critical reflective practice in social ecological systems (SES) research. Following, we introduce critical reflection and the importance of thinking critically in social ecological research as a foundation for the primary contribution of the book, adaptive doing. We then draw on literature from interdisciplinary studies to consider how we can overcome disciplinary divides. We follow this with a consideration of the assumptions related to viewing reality. We reflect on the way we bring order to relationships between humans and nature, with consideration given to determinism, structure, and function, as well as the many different views of time. In response to these challenges with SES research, we turn to an epistemology of practice, along with its ramifications for social and experiential learning, as a foundation for adaptive doing, which is presented in Chapter 4.

EMBEDDED ASSUMPTIONS IN THE LANGUAGE OF SOCIAL ECOLOGICAL SYSTEMS RESEARCH

Active meaning and assumptions emerge from how we construct and use language. Language is dynamic, and this has consequences for both research and practitioners if clarity and common understanding are not

© The Author(s) 2020
A. Rawluk et al., *Practices in Social Ecological Research*,
https://doi.org/10.1007/978-3-030-31189-6_3

established. Hence, a considered discussion of the language used in SES thinking is mandatory. Presently, diverse language is used to express similar ideas in SES research, and similar language is used to express very different ideas. We begin exploring the need for a deliberate practice in grappling with social ecological challenges that creates space for identifying, clarifying, reflecting on, and reconciling the meaning of words. Further, unpacking assumptions of language and examining multiple meanings for the same or similar terms is a key feature of interdisciplinary practice (Bracken and Oughton 2006)—a foundational aspect of social ecological research.

There are many terms used to describe complex systems that involve social/human and ecological/natural elements. The most common are: human–nature relationships (Head 2016, p. 15; Cooke et al. 2016, p. 833), social and ecological systems (Berkes and Jolly 2001; Walker et al. 2006; Ostrom 2009; Smith and Stirling 2010), and coupled systems (Friis and Nielsen 2017; Spies et al. 2014; Liu 2017). Other terms used to denote such systems are socio-ecological system (Gallopín 1991) or human-environment system (Turner et al. 2003; Scholz 2011). Despite being common terms, there has been little reflection on the significance of having multiple terms. For example, Cumming (2014) reviewed multiple frameworks for analysing SES but did not examine differences in terminology used for these different frameworks. In their review of SES frameworks, Binder et al. (2013) noted similar terms in use and implied that they share a common meaning.

Below we examine the definitions offered by each of the authors cited by Binder et al. (2013) along with additional definitions not considered by Binder et al. We consider these to reveal the different emphases and underlying assumptions, and the need to be explicit and transparent when engaging with terms we think equivalent to SES to describe our chosen system in focus. Articulating the foundations of shared terms is central when multiple disciplinarians from the biophysical and social sciences come together to share the challenges that require collaborative responses. Binder et al. (2013) define a **social ecological system** (SES) as a nested, multilevel system that provides essential services to society such as a supply of food, fibre, energy, and drinking water. This definition is based on the seminal work of Berkes and Folke (1998, pp. 185–186). The definition is linked in Binder et al. (2013) to ecosystem services, which implies an anthropocentric meaning to SES as a service to humans, although this does not appear in the Berkes and

Folke (1998) text. Gallopín (1991) defines a socio-ecological system as: '...any system composed by a societal [or human] subsystem and an ecological [or biophysical] subsystem...' (p. 707) that crosses local and global scales. 'Human', 'social', 'ecological', 'ecosystem', 'non-human', 'more-than-human', 'materials', and 'technology' have distinct implicit meaning, often 'saddled with a separationist view of the human' (Head 2016, p. 21). This limited range of terms undermines and in fact demonstrates the 'impossibility of extracting a human body, let alone intentional mind, from the messy relations of the world' (Haraway 2008, cited in Head 2016). This separationist view also constrains how we can practice in and know these relationships, illustrating the challenges we have in shifting our modes of thought, language, and practice (Head 2016).

Turner et al. (2003) describe **coupled human-environment systems** with a similar separationist assumption.

> ...vulnerable systems with diverse and complex linkages...that are predicated on the synergy between the human and biophysical subsystems as they are affected by processes operating at different spatiotemporal (as well as functional) scales. (p. 8075)

Like Gallopín (1991), the limitations of the English language result in a separation of humans and biophysical subsystems, but there is an attempt to overcome this by emphasising the synergy between them. Both Turner et al. (2003) and Gallopín (1991) consider scale, although Turner et al. (2003) include a consideration of time and function, as well as space. Social systems, ecological systems, and the synergy between social and ecological systems potentially operate at different dimensions of time and space (Holling et al. 2002). We return to a discussion of time in SES research and implications for practice later in this chapter.

Other definitions of SES (or similar terms) emphasise the separation of system components, relationships, or the scale of interaction. Similar to Gallopín (1991), Briassoulis (2017) in describing **interlinked human-environment systems**, focuses on the components that make up the system—components, individuals, self-interested agents, and formal and informal institutions. Whereas, Liu (2017) in describing various types of coupled system changes the scale of focus: intra-coupling is about the interactions between human–nature, peri-coupling about adjacent coupled human and natural systems, and tele-coupling concentrates on interactions between distant coupled human and natural systems. These examples are from natural science research, but similar patterns are seen in social science. Smith and Stirling (2010) stress the importance of scalar aspects like change in states over time, interacting scales and levels, and rootedness in a particular context, in defining social ecological systems. Dorward (2014) is more concerned with a self-regulating relationship that maintains negative entropy in describing **livelisystems**, and Lazarus (2017) also concentrates on relationships describing **anthromes** as being unique because of their human-shaped, mutually responsive interactions. These diverse definitions and their underpinnings lead to very different implications for how their users attempt to grapple with and navigate social ecological challenges, that also then have effects on the methods chosen, how challenges are described, and ultimately the recommendations in relation to managing SES.

Other language assumes less separation of the social and ecological in SES. For example, **human-environment systems** suggest holistic consideration of humans and the environment:

> ...coupled, inextricably intertwined systems which are composed of a human-environment complementarity, [wherein] two sets where one contains all the elements that are not an element of the other. (Scholz 2011, p. 31)

Scholz (2011) is the most explicit about his underlying assumptions, and that he is taking a realist, objective perspective, which assumes that the environmental and human systems 'are disjointed but inextricably coupled systems that cannot be well comprehend when considered singularly or disconnectedly' (p. 10). Yet, while the language used in this definition articulates human-environment systems as inextricably coupled systems, the associated framework analyses human and environmental systems separately. There is a common disconnect between how systems are described and how these same systems are studied.

The idea of complexity generally underpins the various definitions of SES, yet this assumption does not always translate into practice. Complex systems and SES are conceptualised as multilevel, expressing emergence, and self-organisation (Capra and Luisi 2014; Briassoulis 2015); as existing in continuous cycles of feedback, in which different parts are highly interdependent and as a whole are co-evolving (Berkes et al. 2003); and as being comprised of many interacting parts and drivers, which cannot be individually distinguished, which result in non-linear effects in the system (Dunn et al. 2017). Such dynamic systems cannot be controlled and problems that exist within them cannot be solved permanently (Dunn et al. 2017). However, the research practices applied to these systems typically break the system into parts and ultimately do not engage with the complexity that leads to SES being ground-breaking. Berkes et al. (2003) acknowledge the delineation between social and natural systems as artificial and arbitrary, yet their discussion and analysis engage with social and ecological systems separately—perhaps unconsciously reflecting the inherent separation between the social and ecological. Turner et al. (2003) engage with complexity in framing human-environment (social ecological) systems, yet discuss human and biophysical subsystems, implying reducible parts of a larger connected system. On one hand, SES are described as complex and their properties emergent, and on the other, the properties are separated and the system reduced to simply parts—which appear to be incompatible lenses.

In this book, we utilise the term **social ecological system** (SES). We intentionally write it without the common nomenclature of a hyphen between the first two words (i.e., social-ecological system). We do this deliberately to give traditional practitioners in the field pause for thought, to disrupt embedded assumptions. The dropped hyphen encourages readers to reflect and to see the term differently. Mary Daly (1978) did the same for words like gyn/ecology in feminist discourse. We wish to show how by examining the construction of what might be perceived as an everyday word in SES thinking, we can think more critically about the meaning of a word, and clarify its meaning and its use. To be explicit, in this book we conceptualise social ecological systems as an intimately entangled and irreducible complex adaptive system that is a co-created phenomenon in time and space, not forced into a binary with a hyphen. For brevity we will use the acronym SES, but we ask that you bear with the clumsiness of the language available and our efforts to emphasise the importance of articulating meaning.

The power of language is always that it can unite understanding as well as divide. Each discipline can draw on other disciplines to create word bridges that better imagine how social ecological thinking could liberate or transform individual discipline practices. Others have found that co-developing models and concepts creates a mutual language for solving problems in SES research (Angelstam et al. 2013). Ideas generated by phrases like 'dynamic systems', words like 'certainty' and 'instability', and concepts like 'emergence' and 'self-organisation' present opportunities to find relevance across the disciplinary divides and to expand our understanding of SES.

EXAMPLES OF THE DIFFICULTY WITH LANGUAGE IN PRACTICE: NON-LINEARITY

To highlight the multiple meanings of a particular term, we draw on the commonly used term 'non-linearity'. Non-linearity is included as a property of coupled human and natural systems (Liu et al. 2007; Friis and Nielsen 2017; Turner et al. 2003), social-ecological systems (Gallopín et al. 2001; Anderies et al. 2006; Ostrom 2009), socioecological assemblages (Briassoulis 2017), livelisystems (Dorward 2014), and anthromes (Lazarus 2017), yet the meaning attributed to non-linearity differs across this literature having implications for how we understand SES and our practices in engaging with them. Drawing on the illustrative case study of this book, *Tarerer*/Kelly Swamp, in the paragraphs that follow, we briefly outline how different interpretations of non-linearity contribute different insights into understanding SES, and how these multiple meanings associated with language might result in different practices and approaches to management.

The first meaning of non-linearity is as a mathematical construct that describes a relationship between the inputs and the outputs of the SES (Gallopín et al. 2001). In this meaning, the outputs are disproportionate to the inputs and the resultant output may be counter-intuitive and a source of great uncertainty. For example, at *Tarerer*/Kelly Swamp a small alteration to an aspect of the system, like the gradual and small annual sea-level rise that has occurred along the south-west coast near Warrnambool to the east of *Tarerer*/Kelly Swamp has contributed to disproportionately larger blow outs in the sand dunes following storm surges (EPA 2011). This interpretation of non-linearity focuses on identifying the important inputs and outputs and monitoring these to be able to detect changes in structures.

An alternative meaning of non-linearity is employed by Anderies et al. (2006), and Berkes and Folke (1998), who emphasise the emergent patterns exhibited in systems—with multiple stable domains or multi-equilibriums. *Tarerer*/Kelly Swamp has at least two stable domains. Shortly after a flood where fresh water is in abundance, organisms that are dependent on fresh water utilise this site, including cows and farmers.

Later in summer when significant evaporation and groundwater pumping for irrigation has occurred, the wetland stabilises as a more saline habitat. *Tarerer*/Kelly Swamp exists with water in it and there is a causal or linear link between the presence of water and certain species inhabiting the site. In the absence of water, life continues for many other, or most of the locally recorded species, indicating they are not just dependent on the surface water for their existence. They have seasonal, emergent life cycles and habitats that are part of the more complex ecosystem. In managing this SES, statistical and the emergent views of non-linearity might focus upon seasonal relationships. By recognising these two different meanings of non-linearity, there is a space that opens to consider alternative ways of viewing the SES.

An assemblage view, which draws on Deleuzean philosophy, presents yet another understanding of non-linearity—where the assemblage or system exists as multiplicities, is provisional, and yet a seamless whole. In other words, the elements of the assemblage may reorder to fulfil a purpose and yet the whole appears the same and complete (Briassoulis 2015). Here, the emphasis of non-linearity is on the ordering and relationships between the elements—the structure and function may change but the system remains the same. This is evident in *Tarerer*/Kelly Swamp, when superficially the level of water visible over the surface appears stable for long periods, but over time, the source of the water at the surface is likely to differ and its salinity will change. In winter through into spring, when rainfall is more abundant, the water will be predominantly fresh water that has flowed down the Merri River or filled from underground seeps in the landscape. In summer and into autumn this is likely to shift to spring-fed water running through aquifers and stored water from the dunes seeping back into the land. Depending on the senses and measures used, we may not detect certain changes over time, that superficially appear the same, but their structure or function may differ. Including such a non-linear view when engaging with SES may encourage us to consider other senses and measures, keeping open alternative options for thinking about and managing such systems.

A common theme across these views of non-linearity is the challenge in predicting the behaviour of systems that exhibit this property, no matter the scale (Gallopín et al. 2001). Smith and Stirling (2010) have suggested it would be helpful to acknowledge the contextual, contingent and multiplicities associated with the non-linearity of SES by shifting governance practices to emphasise learning, experimentation, and iteration. Greater transparency in the meanings attributed to properties associated with SES, like non-linearity, may also assist to open up alternative ways of understanding and acting to manage these complex systems. Recognising different views of non-linearity and the inherent uncertainty of SES is a way of ordering that assists us to understand the SES we study in a rational way, and it may predict the kinds of stories that we can tell about an SES. In this book, we acknowledge that one way of seeing SES is as assemblages (Cooke et al. 2016) that are constantly in the making through the active cohabitation of humans and non-humans. We consider multiple views of non-linearity when we further examine *Tarerer*/Kelly Swamp in Chapter 5, as we wish to keep the story about the SES as open and broad as possible, with many future story options.

CRITICAL REFLECTION FOR SOCIAL ECOLOGICAL SYSTEMS RESEARCH

Critical reflection is a key tool for engaging with the uncertainty in systems, as well as the multiple assumptions and understandings involved, because it assists in crossing boundaries of knowledge and ethics, and can facilitate consideration of alternative futures. We introduce **critical reflection** as central to this book. It assists change to one's understanding of self in research and of a system generally. Our use of critical reflection is informed by the scholarship on critical pedagogy and adult education (e.g. Freire 1970; Mezirow 1998), wherein people deliberately disrupt and challenge assumptions, language, and meaning, and consider alternatives.

We draw on consciencisation or critical consciousness (Freire 1970) to inform our framing of critical reflection. Consciencisation, has emancipatory origins that strive to develop an awareness of social structures and how they can be deconstructed, thereby helping to ensure reflections are 'critical'. Consciencisation is the deliberate act of becoming conscious of how we are educated and what we are taught, and brings into focus the kinds of questions we need to ask about the purpose of learning and the accumulation of knowledge (Freire 1970). Consequently, consciencisation assists with the process of examining assumptions, giving rise to new practices.

In the context of SES research, if we are to envision and construe actual sustainable futures, we must first understand what brought us here, where the roots of the problems lie, and how the sustainability discourse and framework tackle—or fail to tackle—the underlying challenges. Taking this critical perspective intentionally politicises sustainability and SES thinking (Ferreira 2017). In SES research, this may include assumptions about language, as discussed in the previous section. It also requires questioning of assumptions related to the integration of disciplines, how we order knowledge, and our practices. We start with interdisciplinarity, as a core approach in SES research and management, and then turn our attention to ordering of knowledge and practices.

A New Chapter of Social Ecological Interdisciplinarity

Involvement in SES research and practice means acknowledging and taking on the challenges and benefits that come with a diversity of disciplines, cultures, knowledges, and practices. To date, this diversity has been expressed mainly as a challenge, and critical engagement with the assumptions of reality, knowledge, and good practice brought to any form of SES management is a key step in finding creative and adaptive ways forward (Phoenix et al. 2013). For example, Briassoullis (2017) argues that there are different knowledge cultures that underpin three different framings of SES: a singular, whole system as represented in panarchy (Gunderson and Holling 2002); networks (Anderson and McFarlane 2011; Anderson et al. 2012); and assemblages (DeLanda 2006). Cumming (2014, p. 19) echoes this challenge, stating that 'one of the weakest links in the study of SES remains the one between different epistemologies [ways of understanding knowledge]'. While Briassoulis (2017) argues that these different knowledge cultures are incommensurable in social ecological research, in this book we take a more optimistic view and engage directly with the cognitive challenge of integration (Luthe 2017). As authors from multiple knowledge cultures, we suggest that transparency, a willingness to be challenged in our worldviews, and to find common pathways and approaches towards integrated practices is central to the future of social ecological research.

Navigating the changing boundaries of disciplines and their involvement in interdisciplinary work has also been viewed as a challenge. Disciplines are continuously evolving, despite a frequent assumption that they are static, fixed, and fully formed. Disciplines, such as sociology or ecology, are maintained by: turning inward and strengthening boundaries; forming strategic alliances with stronger disciplines; and reconstituting the discipline in a newer and larger field of study (Krishnan 2009, p. 47). However, Popper (1963) wrote that 'we are not students of some subject matter, but students of problems' and his intention was to facilitate the solving of a problem rather than the reinforcing of discipline boundaries. Angelstam et al. (2013), and the authors of this book echo Popper (1963) in their call for research that blurs disciplinary boundaries to be considered as an applied practice, evolving from current problems of the world, not attached to pre-established method or design.

Despite the interest in interdisciplinary approaches to complex social ecological issues, there have been ongoing challenges associated with the process of interdisciplinary practice. While researchers arrive at social ecological challenges open-mindedly and with great intentions of creating change, we so often return to our disciplinary assumptions. This often leads to multidisciplinary research, as opposed to integrated knowing as fostered in interdisciplinarity. For example, the Australian Farms, Rivers and Markets project (2012) claimed its interdisciplinarity as core to reaching useful outcomes, as it investigated the complex relationships involved in water scarcity across Australian production landscapes. The research brief included the objective of integrating the approaches and findings across the many disciplines involved in the research (Ayre and Nettle 2015). However, the authors report the difficulties that ensued in attempting to fulfil this aspect of the research, noting how time and funding constraints were used by discipline-based participants to resist forums designed to create or reflect on interdisciplinarity and only partial achievement of the multidisciplinary outcomes.

In SES research, both disciplines and interdisciplinarity are important. Wernli and Darbellay (2016) argue that interdisciplinary approaches can complement disciplinary research, and in effect good disciplinary knowledge underpins interdisciplinary research. This may be addressed with a top-down directive to use interdisciplinary approaches, and a bottom-up interdisciplinary approach that extends or creates new knowledge. This top-down and bottom-up reality is framed as a 'virtuous circle' between disciplinarity and interdisciplinarity (Wernli and Darbellay 2016, p. 4). Participants involved in interdisciplinary projects, like SES research and practice, must be willing to engage with this virtuous circle and a process of critical reflection, testing assumptions and a preparedness to give the time to negotiate and learn.

A key knowledge-related (epistemological) challenge in interdisciplinary SES research is the imperative to bring together knowledge and practices from different disciplines of the natural and social sciences. While much research has focused on a natural science approach to examining SES, underpinned by a particular understanding of reality, knowledge, and practice, various disciplines in the social sciences provide useful insights. For example, there is importance and value in qualitative and constructivist research.

> Process-oriented, and... social constructivist research approaches present the means to tackle the challenge of setting system boundaries... [because] systems are intellectual constructs rather than predefined entities. (Friis and Nielsen 2017, p. 12)

Moore (2015) re-conceptualises the social ecological in a more dynamic and interdisciplinary way, in which our practices evolve methods that reflect the social ecological imagination through our actions (reactions) as humans-in-nature or humanity/environments. In moving into a new chapter of interdisciplinarity in social ecological research, we need to leave behind a siloed disciplinary understanding of phenomena. There is a diversity of research practices in SES studies that can create space for multiple understandings and imaginings of the social ecological while examining the system, and its associated challenges. Restricting to only one way of conceptualising the social ecological limits the possibilities for collaborative practice that works across multiple understandings and experiences to address the challenges of the twenty-first century, such as the entanglement of social ecological justice (Collard et al. 2018).

The second knowledge-related challenge is between Cartesian and complexity science. Despite the association of SES with complexity, there is still a common assumption that SES are reducible and quantifiable. Building on the writing of Capra and Luisi (2014), the characteristics of complex systems thinking are: a shift of perspective from the parts to the whole; inherent multi-disciplinarity; the study of objects becomes the recognition of relationships; a turn from measuring to mapping patterns; a shift in focus from predominantly on quantity to an emphasis on qualitative analysis; a change in emphasis from the study of structures to the recognition of processes; a reorientation from objective to epistemic science that emphasises the importance of questioning; and change from Cartesian certainty to approximate knowledge. In the context of *Tarerer*/Kelly Swamp, we can study individual aspects of the Swamp species, or individual species without necessarily considering their connectivity to the wider conditions affecting the Swamp, like the agricultural management of the surrounding landscape or the likelihood of *El Niño* weather patterns changing water levels and heat tolerances. Our monitoring may reflect all of these conditions in numbers of species accounted for, or, it may provide only partial representation. A systems approach locates individual studies more forcefully within a larger framing of the Swamp or the landscape or the region. Systems thinking does not negate the micro examples but builds strongly on the ideas of relational and connected knowing. This assists us to understand *Tarerer*/Kelly Swamp in the context of the wetlands and drains that persist across the western plains. The social ecological processes that have alienated water movement from pre-European times to the present provide geological and paleological patterns that can be researched and contribute to our management of particular sites. We draw on our conceptualisation of SES as underpinned by a complex systems-thinking approach in our examination of the Swamp SES in Chapter 5.

Taking an interdiscplinary approach is an inherent aspect of SES research and practice. Here we argue that assumptions of culture and reality, how we do interdisciplinarity, as well as how we bring knowledges and practices together must be tested and engaged with actively. We also argue that complex systems thinking and critical reflection can assist our efforts. Further challenges that must be examined include how we order knowledge and the attention we bring to practice.

Bringing Critical Reflection to the Ordering of Knowledge

In this section, we examine the impacts that the different ways we order knowledge can have on the way we understand SES and the practices we bring to bear. We start with considering the effects of uncertainty; we then reflect on the relationship between knowledge and practice; lastly, we consider our relationship with 'nature' and how an emphasis on relationships, more generally, can change how we order, and our ethical consideration of the unseen or unheard.

Framing the SES within complexity thinking involves an acknowledgement that uncertainty is ever-present. Recognising uncertainty changes the way we order the world around us. Capra and Luisi explain how order and disorder occur simultaneously, meaning that as Prigogine (1989 in Capra and Luisi 2014) argued, at critical points of instability, order can emerge. A simplified system of ordering, such as the order that a hierarchy imposes, does not assist us to imagine the self-organisation and emergence of new ways of ordering that arise in dynamic systems. This is because of the inherent instability within these systems. E.O. Wilson framed complexity as creating unity through recognising (unexpected) connections. In other words, Wilson argued that those endeavouring to know systems would encounter surprise (Wilson 1998). Efforts to apply an order can be a problem because they can inappropriately simplify and give a false understanding of the SES system, limiting the possibility of seeing alternative arrangements of ideas and system elements, such as through inter-species associations.

Ordering tends to assume that knowledge is separate, pre-existing and 'applied' to practice (what we do), whereas we argue that knowledge and practice are co-produced, forming each other as each evolves, as described by Cook and Wagenaar (2012). Efforts to extract linear evidence of causation is an issue for a number of reasons. It de-emphasises and perhaps hides the less seen connections. It often does not acknowledge or give recognition to the non-human and unseen connections that may act, have agency or engage. It is also a reflection of our limited vision, rather than a reflection of the entirety of the system.

Whether we view ourselves as part of or separate from nature also impacts on how we order knowledge. Ordering is not inherently bad. It can be helpful because it assists researchers to understand phenomenon by identifying the components. In Chapter 2, in describing *Tarerer/Kelly Swamp* we first outlined the biophysical context and then the social context. This is a common form of ordering information within ecological studies of SES. By imposing such an order, where the biophysical is separated from the social, we demonstrate an assumption of this separation and potentially reinforce binaries that set humans outside of nature, and this may undermine efforts to recognise the connections and interdependence between all organisms and the complexity of the systems we inhabit. By separating and ordering the components of the SES in this way, our understanding of the complexity and the relations between these elements are also potentially limited. We need to critically reflect on what is ordered and how we order it particularly because our efforts to create boundaries and order tell us little about the kinds of social ecological relationships we need to explore or the kinds of entanglements we believe are important to effect change.

Relational ways of knowing bring a different order to how we understand an SES. Focusing on relationality means that we attempt to accept non-linear, non-hierarchical ways of understanding how elements are connected, and it places an emphasis on qualitative analysis that documents the processes of engagement, rather than attempting to extract linear evidence of causation. Things may indeed be causal, reactive, and linear, but often this is about what we see—the tangible connections—like the bird flying to a particular vegetation type. A relational approach brings order through identification of relationships that may not have been seen but seem probable. By emphasising relationships as a form of ordering, we intend that these relational assemblies be understood as dynamic and therefore part of other less seen connections. The purpose of reconceptualising the SES is to constantly require an awareness of the unseen and to pay attention to decisions we make as a consequence of that novel recognition. Ideally, the non-human action, its agency, and our mutual engagement as humans with the non-human will facilitate an awareness of how we understand relationships among elements and ideas in the SES.

Relationships in social ecological systems are often understood in partial ways. Ecosystem services frameworks consider one type of relationship, the range of benefits that ecosystems provide for human society (Chaudhary et al. 2015). Studies of connectedness to nature and pro-environmental behaviours in psychology explore positive relationships that individual people have with their surroundings, while ecologists study the negative impacts that aggregate human activities have for ecosystem components such as species habitat. While studies within disciplines deepen understanding of certain relationships, the multitude of relationships are rarely observed together. By bringing together the many relationships, as a form of ordering, the mechanisms by which relationships are initiated, held together or supported (organisations, power, trust, credibility, for example in society; and habitat structure and function in ecological contexts) are made more evident.

However, ordering may also result in a number of unquestioned impacts on the boundaries of the subsystems, such as what is observed and what is not, who's voices count (or not), either explicitly or implicitly in the research (Law 1999; Mol 1999; Blaser 2013), as well as reinforcing how we perceive ourselves in relation to nature. We have a moral and an ethical imperative to consider our practices and the way we order information, as well as its impact on who's knowledge and experiences are privileged and who's knowledge and experiences are silenced.

As an extension to this discussion of ordering and its impacts on how we understand SES and the practices we bring to SES management, we discuss the issues presented by a deterministic ordering, the strengths and weakness of structure and function ordering, and the gaps in the ordering of time. Subsequently, we turn to consider how an epistemology of practice relates to learning, and how this offers helpful responses to the challenges we articulate in this chapter.

Turning Away from the Assumption of Social and Ecological Determinism

In SES studies, determinism appears in the form of ecological and social determinism. Ecological determinism assumes that geography or environmental conditions determine the organisms that occur in a particular place (O'Malley 2008), while social determinism suggests that interactions and societal influence determine behaviour more than, for example, biological reasons (see Durkheim 1982). Deterministic thinking shapes, and potentially limits, the future imaginings of a particular context. The Anthropocene is arguably a new geological age in which human activity dominates the environment. Climate change is usually given as an example of this sort of human-made activity. In their 2009 paper 'Planetary boundaries: exploring the safe operating space for humanity', Rockstrom et al. (2009) acknowledge that 'there are large uncertainties and knowledge gaps', and assert the ecological or biophysical determinants of future Earth already exist, are understood and defined. This is an attractive moment of social determinism (humans know and can predict climate overshoot). It may seem that if we stay within these defined but also undefined boundaries, we will be safe. But we know that this kind of boundary is also an example of ecological determinism. It is what we have been contemplating with climate change temperatures—and clearly knowing that we want to stay below two degrees of warming has not led to significant changes in useful policy or to indications that we can manage to control Earth's warming and maintain 'safe operating spaces' (Aton 2017). As White et al. (2017) would argue, these boundaries tell us little about the kinds of socio-ecological relationships we need to explore or the kinds of entanglements we believe are important to effect change in the way we practice our research and develop and enact our methods so that we can overcome the limitations of our siloed social and ecological constraints. As well, the lack of interaction, integration, or inter-penetration of the social ecological negatively consolidates the boundaries that science is arguing over. These are seen as believable only to scientists, and citizens, by and large, do not feel that they have to engage with these ideas (Climate Reality Project 2017; Dixon 2017). SES research and practice would benefit from critical reflection on how we order knowledge, placing a stronger emphasis on the seen and unseen relationships, and acknowledging determinism when it occurs.

Structure and Function

A further example of the difficulty with ordering is apparent in the use of structure and function in characterising systems. An ecosystem's structure is the physical features and the organisms within it. A swamp has abiotic structures such as soils, water, and solar radiation, and it has biotic structures consisting of vegetation and animals (Odum 1975). The function of an ecosystem is its processes, particularly those that have a role in sustaining life. Studies of function may include the flows of materials and the movements of organisms. In a swamp, cycles of wetting and drying contribute to functions such as nutrient and habitat provision for different species, as shown for *Tarerer*/Kelly Swamp in Chapter 2. Structure and function have implications for the stability of ecosystems, for example a diversity of vegetation species with different rooting depths enhances productivity and is thought to also increase the range of functions that may operate. Ecosystem stability is particularly enhanced by 'response diversity', where species that have a similar function in the ecosystem have different responses to disturbance (Walker et al. 2006). Furthermore, it is not uncommon in ecology to describe functional roles, such as predators or herbivores, shaping the structure of communities; i.e., function leads to structure (Begon et al. 1990), which serves as an attempt to find an underlying causative reality, beneath observed phenomenon. Such labelling of predators may be based on limited observation leading to determinist expectations.

The same terms have been used to describe social systems, but this has at times been problematic. An early approach to the social in SES thinking used 'structural functionalism' that has origins in biological sciences and has been criticised for not grappling with issues of power and social change (Hatt 2013). Social diversity also tends to be underplayed in SES thinking (Fabinyi et al. 2014). A use of the term structure that better reflects the multiplicity of ideas and things it covers, is for laws, customs, and divisions such as those of class, gender, race, and ideology. These social structures, such as class, are associated with processes that serve to maintain them. In this case, it is an argument that structure leads to function. Reification is where parts of life are abstracted or objectified, such that work and other interactions with capitalist economies become objective parts of people's lives, separated from the subjectivity of human experience. The objectified activities, though constantly reproduced, appear to be stable, with rules that are difficult to change, such as those of the market. In this way, reification and other social processes confer an apparent stability that can limit responsible action in the face of crises such as climate change (Leduc Browne 2018). However, functionalists in the social sciences argue that an underlying social rationality causes individuals to act to fulfil society's needs (Capra and Luisi 2014), i.e., function leads to structures. Later social science researchers have argued that the relationship between structure and function is cyclical (Giddens cited by Capra and Luisi 2014, p. 300), with 'social structures being both the precondition and the unintended outcome of people's agency'. Capra and Luisi (2014) suggest in integrating these perspectives that process is the link between structure and function, each offers different but an inseparable perspective on the phenomenon of life. Analysing social ecological structure and function requires remaining alert to the different ways these terms are used and the assumptions they reflect.

In Chapters 4 and 5, we return to interrogate these assumptions of language, ordering, and structure and function, through the process of 'adaptive doing'. Within the adaptive doing process, we draw upon the 4 Is reframing tool (Beilin and Bender 2011), which assists with critical reflection on these assumptions. Below we consider the common assumptions related to time and their implications for SES research and practice.

Attending to Time in Social Ecological Research

It is not uncommon in social ecological research for time to appear 'as a location, and the phenomenon (to be) examined as a series of snapshots, with time locating the phenomenon in two or more places' (Cooke et al. 2016, p. 832). Such a view of an SES can be seen in the management plan written for *Tarerer*/Kelly Swamp and surrounds in 1993. This recognises past actions like the creation of a permanent coastal reserve for the prevention of irruption of sand in 1873 that is now managed under the Crown Lands Act (1978). This is primarily an effort to document the state of the biophysical elements (native and pest flora and fauna, water management, topography, fire management) and cultural values of the SES at that time. From this static 'snapshot', suggestions are made about what management should be taken into the future. It serves as a picture of the state of the system, which makes it hard to recognise that the system is undergoing constant change. Like many studies, it includes an implied recognition of time passing and change through the ordering of information into past, present, and future. Temporal separation between cause and effect often contributes to the intractable nature of problems (Meadows 1982, cited by Allison and Hobbs 2004). Time is simply another form of ordering, often sequenced as if distinct, as in past, present, and future.

One heuristic for visualising time and continuous change in SES research and resilience thinking is the adaptive cycle. There are multiple interpretations of the adaptive cycle heuristic, its relationship with time, and its use, such as to linearly explore past changes (e.g., Sinclair et al. 2014), future trajectories (e.g., Walker et al. 2009), long-term slow change (e.g., Redman and Kinzig 2003), and short-term change (e.g., Lyon and Parkins 2013). In addition, there has been recognition that each SES is unique in the amount of time it remains in any one of the four phases (Walker and Salt 2006, 2012). Boundaries between the four phases, however, must be negotiated, are highly uncertain (Cooke et al. 2016), and it remains a challenge to reconcile causal relationships across time. Accepting that an SES may be recovering from disturbance, growing, sustaining equilibrium, or experiencing collapse (Kharazzi et al. 2016) is not unlike the recognition that any particular system, like a swamp, may have multiple stable system states, multiple future realities and that SES are patchy in their relationship with time.

There is recognition that some elements in an SES operate at different timescales to others, such as water table depth, native vegetation cover, riverine ecosystem condition, and values which Walker et al. (2009) identify as controlling (slow) variables. This recognises multiple timelines and implies multiple realities each with their own time relationship.

There has been less acknowledgement that subsystems, like the indigenous vegetation community in a water catchment might be in a different phase in the adaptive cycle (e.g., collapse) than the phase attributed to the rest of the system (growth or conservation). How we view SES in time, as snapshots that represent a single linear reality, or many possible pathways, changes how we understand the SES, the possible futures, and the practices we implement to shape the outcomes. Keeping multiple timelines and multiple futures open changes how we conceptualise SES and the practices we bring. Drawing on the work of Blaser (2013) in anthropology, we can see each of these multiple future states as different ontologies, or worldings that are complexly entangled, taking place in the same spatiotemporal location but not always interfering with each other.

In writings on time by Indigenous scholars, as well as non-Indigenous scholarly representations of Indigenous knowledge and practices on time that are included in SES literature, time is expressed in ways that remove the usual boundaries and ordering similarly to 'worldings'. Vanessa Watts (2013) depicts the world (tangible and intangible, human and non-human) as cyclical, describing continuous change that transcends past, present, and future. Blaser (2013) describes all parts of a system as expected to change with knowledge systems and practices returning to past knowledge and practices that have an inherent wisdom but were rejected in the project towards modernity. These ideas are captured in the *Quechua* word *pachakuti, where pach* means (world, time and space, or state of being) and *kuti* means (change, turn, or something that comes back on itself) (Blaser 2013). Focusing on time as cyclical in an irreducible SES, would have distinct implications for practice. Such rethinking of how we see time and the different views of SES signals a struggle to define the limits of the political and the process of change.

Nevertheless, a Western scientific notion of time has predominated much SES research. Walker et al. (2006) recognised that in ecosystems, the variables that control the shift from one regime in a system to another tend to change slowly, while managers frequently focus on fast variables, suggesting an existing mismatch between the important variables and where management attention is focused. At *Tarerer/* Kelly Swamp, the amount of salt in the water is an ongoing concern for farmers and ecologists alike, although there is no documented case of 'salinity' issues at this time. Salt concentration in the water is a slow variable, but management is focused on vegetation type, cover, and protection, all of which change over short time scales. Identification of multiple timelines, how they interact and how our practices can incorporate multiple understandings of time is needed. Gallopín et al. (2001) observed that in experimental science studies there is a tendency to have a narrow enough focus in order to pose hypotheses, collect data and design critical tests, so as to reject invalid hypotheses, but that this results in the chosen scale being small in space and short in time. This may give the impression that the timeline for the study is neutral, but if time is seen as outside of the experiment with the observer, then there is a power inequity. The timeframe that is used to understand SES can therefore have implications for what we can know about SES and can inhibit how we manage SES if there is a mismatch between the issue and how we seek to understand it.

When undesirable properties of an SES are delayed in their expression (time-lags), or become locked-in through path-dependency (where past events restrict the choices for current decisions), or experience legacy effects (where human disturbance in the past is the cause of current and ongoing environmental changes), then additional challenges in understanding and managing SES are introduced. Lazarus (2017) notes that in such circumstances it becomes difficult to disentangle drivers from stressors, and to classify the type of system operating. Here there is a mismatch in our abilities to detect or respond to changes in a system because of the speed at which they are occurring, or because we perceive them to be irreversible or effectively irreversible (Walker et al. 2006). Thus, historical changes and slow responding systems present time challenges. These limit our ability to understand what is happening in a system, and we would benefit from expanding our methodological and epistemological processes to engage more productively with time.

There are direct implications of how we understand time in the possibilities and approaches for SES management. Management of SES may be evaluated over different time scales, leading to different conclusions about the sustainability and resilience of SES and their management (Lazarus 2017). At each unique moment in time and space, there will be potentially very different 'best' management responses. However, time and how it is conceptualised and understood is generally only indirectly referenced in resource management (Gallopín et al. 2001). While Gallopín et al. (2001) suggest that how time is conceptualised needs to be broadened to accommodate the need for intra-generational as well as inter-generational equity considerations, we suggest that there is a lack of guidance on how to collaboratively understand and engage with time and even multiple interacting forms of time. So, there is a need to reconsider how we order chronological time which we presently describe as past, present, and future, and to bring an ethical lens that considers both the current generation and future generations that does not privilege either.

Cook and Wagenaar (2012) discuss a different conceptualisation of time that is tied more closely to our focus on practice—the relationship of time through an idea called the 'eternally unfolding present' (EUP). They describe the EUP as a temporal zone where the past, present and future are all interwoven through an infinite span of present moments, shaping and being shaped by each other, and by extension, all the social ecological elements and ideas in the SES. Changing environmental conditions and changing knowledge and theory, result in the frame and the system being in a constant state of response and change. Time in the EUP has an elasticity and is dependent on the emergence within the SES—the SES of this moment is not the same as the SES in the next moment. It seems that there is a normative use of time in SES research which suggests chronology. Such an ordering is not the same for all the parts that are part of an SES—soil has a different timeline to climate, law, and dairy cows. But in general, we know that what is meant is a chronological ordering. What we want to emphasise is that this covers a lot of assumptions about how everything else is ordered.

The idea of the EUP comes from the Japanese philosopher Nishida who holds that the past is accessible only in the present, where we constantly assemble and reassemble it as we move through time as 'on-going business' (Cook and Wagenaar 2012). Memories, then, are reconstructions in the present of past events. Likewise, the future remains a feature of our present imaginings, such that we exist in an eternally unfolding present (Cook and Wagenaar 2012). Time is not an objective aspect of SES change and management and its subjectivity is demonstrable if we consider how experiences such as memory shape and are shaped by an understanding of time. Collective and social memory shape what is remembered about the past, as well as what is forgotten (Keightley and Pickering 2012). Memory and the perception of time is thus an expression of complex and intimate power dynamics (Fine and McDonnell 2007). How a present moment in an SES is understood is an integration of past experiences, which often guide future imaginaries and decision-making (Rawluk and Curtis 2016). Watts (2013) reminds us that memory and the action of remembering is not 'a question of going backwards, for this implies there is a static place to return to' instead, the *Haudenosaunee* and *Anishnaabe* Indigenous peoples she writes of, draw attention to the importance of listening and then acting. Little attention has been given to practices for understanding, integrating, and reconciling multiple and contrasting memories.

Cook and Wagenaar (2012) also suggest that there is a need for a reconceptualisation of the irreducibility of knowledge and practice, rather than more knowledge, because there is a relationship between learning and time and how we order what we learn. Our learning is subject to reorganisation over time and different aspects of our understanding develop at each encounter with what begins as the same knowledge but upon revisiting may lead to different understanding. We are suggesting in this book that we need to bring this idea of non-linear and multiple times into the reconceptualisation of our methods to at least give space for time.

Openness to an ontological repositioning of time, what it is and how we engage with it, is a starting point to orienting SES research to practice. Each unquestioned aspect of time potentially shadows important elements of the SES being studied. To make new possibilities, Datta (2015) argues for a relational ontology that gives validity to multiple forms of knowledge, Indigenous and Western scientific, and a requirement that our thinking patterns do not position us as outside of societal realities. In Chapter 5, we draw on assemblage as a reframing tool to engage with relationality and the EUP as a reframing tool to engage with the non-linearity of time and timelessness for seeing SES differently. Seeing time as dynamic and with multiple interpretations keeps multiple futures open, recognising multiple views, knowledges, and practices. It also means understanding that we 'construct' the future, in the present.

Centring on Learning and Practice: Turning to an Epistemology of Practice

Much SES literature reflects a technically rational view of knowledge, in which what is known or observed is shaped by theory, and expert disciplinary knowledge is 'applied' to practice. For example, monitoring of the condition of a swamp is understood as a task for experts who hold and apply appropriate knowledge. In our initial description of *Tarerer/* Kelly Swamp in Chapter 2, knowledge is presented in the received way as information from various published sources, that appears largely to have been gained through discrete disciplinary activities. There has also been a strand within SES thinking to date on learning as a way of adapting within complex and dynamic SES. Learning is claimed to be necessary for working through change and uncertainty, but often with little attention to the processes by which learning occurs in SES (Armitage et al. 2008). Our initial description of the Swamp lacks any discussion of opportunities for learning to respond to complex changes, such as might be driven by shifts in climate, or in markets for agricultural products. Armitage et al. (2008), among others, point to many concepts of learning in which critical reflection is implicit and necessary, and which have the potential to become more prominent in SES thinking. A major contribution of studies on learning has been to identify ways to describe and enhance learning that can assist all kinds of practitioners to engage in shared critical reflection.

There are many conceptualisations of learning, from individual to organisational, which we draw on for social ecological practice. A key distinction is made between single loop, double loop, and triple loop learning. Single and double loop learning were originally described based on observations of staff and managers learning from feedback in organisations (Argyris 1976). Single loop learning incorporates new information only in a way that is consistent with governing frameworks, values, principles, and beliefs, without questioning these. Double loop learning in contrast considers the frameworks, values, and beliefs as information for learning, and opens these up to be changed (Argyris 1976). Triple loop learning involves changes to governance norms, rules, and protocols (Armitage et al. 2008). Learning across the social and ecological parts of a system requires at least double loop learning.

We now take a closer look at how learning occurs in individuals. Learning was originally defined within psychological and phenomenological perspectives with different terms, but with a similar emphasis on learning as an active process in which people interact with the world around them. From a psychological perspective, individual learning is process in which new information is related to multiple linked internal representations of the external world. Cognitive maps (knowledge structures) consist of many concepts including objects, values, attitudes, and links among them that have been formed by past experiences. Once formed, these can be activated by new information encountered in the world, such as observations of landscape elements, or changes in a system. However, limits to human attention mean it can be difficult to attend to new information unless it is considered relevant to existing ideas, and it must be actively integrated into the cognitive map or it will be difficult to draw on in future. This active nature of learning means that new information may not lead to knowledge unless it is salient and its meaning fully considered (Kearney 1994). Cognitive maps include dynamic working models of systems, such as ideas about how social-ecological systems function (Steg et al. 2013). These models of how system changes guide people's actions are conceptual. Although the models used to justify an action are not necessarily consistent with how people behave, to make shifts in understanding of the system and in related actions, people must change these models, a demanding process of double loop learning (Argyris 1976; Fazey et al. 2005).

Phenomenological studies focus not on cognitive structures, but on how learners describe or demonstrate their own experience of learning. A dynamic relationship between the person and the world is emphasised, and learning is considered to be in the changes that occur when a person has new experiences and develops a new understanding of their place in the world (Fazey et al. 2005). Mezirow (1997) describes transformational learning as a process of changing one's frame of reference to one that is more inclusive, self-reflective, and integrative of experience. This is thought to occur through critical reflections on the assumptions on which our interpretations and points of view are based, through four processes: elaborating a point of view by seeking further evidence; establishing new points of view; transforming a point of view through experience; and finally, by transforming habits of mind and frames of reference, for example by opening these to reflection and identifying and removing biases. This awareness and reflection on one's own assumptions is another example of double loop learning, and is indispensable for adapting to change (Mezirow 1997). This also echoes the intentions of Freire's (1970) commitment to consciencisation as a part of educational change.

Consistent with knowledge emerging through practice, individual learning involves emotion and intuition, not only reasoning. Emotions, understood as bodily feelings, and intuitions, such as quick moral judgements, play a critical role in practice in interaction with cognition. Emotions make decisions possible, as is evident from observations of patients with frontal lobe damage who have extreme difficulty deciding what to do. Cognitive processes provide access to the information needed to make a decision, while emotions (bodily feelings), originating from our positive and negative previous experiences, enable us to rapidly test whether information is salient (Saver and Damasio 1991). Similarly, intuitions such as quick moral judgements about what is right or wrong, enable tacit knowledge about what is appropriate to do. Haidt and Joseph (2004) argue that moral intuitions are an innate characteristic of humans, a preparedness to feel approval or disapproval towards patterns of events involving other people. Different cultures build varying moral systems on this foundation of shared intuitive ethics. In these ways, knowledge that emerges from practice is far removed from the received view of separate packages of information that are simply applied.

Experiential and social forms of learning are particularly relevant for understanding how people may encounter and engage with new information within their everyday practices in SES. At its simplest, experiential learning is described as an individual process with stages occurring in an iterative cycle, such as: concrete experience (doing); reflective observation (reviewing); abstract conceptualisation (concluding); and active experimentation (planning) (Kolb 1984). The learning cycle concept has been adapted to describe experiential learning by groups of people engaged in joint activities in the environment (Armitage et al. 2008), which can inform the design of group processes to foster learning. A staged approach might include: engaging in experimental action despite uncertainty; monitoring outcomes to obtain new data to inform future decisions; stimulating discussion and reflections on findings; and incorporating new conceptualisations into future action phases. Such processes are essentially tools for slightly modifying the ongoing flow of practice to foster deliberate critical reflection, and thereby to encourage learning.

The term social learning is used in many different ways. It was originally described as learning by observing other's behaviour, a process that can be enhanced by an interesting model who captures attention and by the learner performing the behaviour under supervision (Bandura 1971). Social learning is also described as occurring through shifts in shared mental maps and models, also termed social representations. These are considered to be socially constructed, and continually re-constructed, through everyday interactions and discourses, such that different social representations of the same object may form through the everyday interactions of people within different social groups. Social learning is also described phenomenologically as learning that occurs through engaging with other people, for example, 'Learning that occurs when people deliberately engage each other, sharing diverse perspectives and experiences to develop a common frame of understanding and basis for joint action' (Schusler et al. 2003, p. 311). Social learning is linked to double loop learning, for example Argyris (1976) describes double loop learning as being fostered by collegial styles of decision-making.

For tackling complex sustainability problems in SES, learning must occur not only through individual experience, as in the learning cycle, or among small groups, as in some descriptions of social learning, but also on a wider scale. Ison et al. (2013) explored many ways that social learning is conceptualised by studying metaphors. Several of these emphasise action within complex systems. Metaphors of social learning as a performance, such as an orchestra, reveal complexity and are underpinned by ideas of praxis (practical action informed by theory). They are consistent with Ison et al.'s argument that social learning can be viewed as a duality, as both an entity (set of governance practices) and a process (processes enacted in social dynamics). Metaphors of social learning as a governance mechanism, such as a conceptual framework, or process, or adaptive mechanism, reveal associations between social learning and ideas about good governance. Social learning as a conceptual framework suggests it can be used to structure or shape thinking and could be interpreted as an epistemology.

Studies of learning in governance suggest experiential social learning occurs through complex and dynamic interconnected networks of people (Pahl-Wostl 2009). Learning in such contexts has been explored in agricultural extension (Engel and Saloman 1997), public administration (Cook and Wagenaar 2012) and SES governance (Pahl-Wostl 2009). Pahl-Wostl (2009) discusses adaptive management at a broad scale, in multilevel governance of the environment in response to climate and other changes. She argues that in SES, changes in governance are social learning processes that occur in and between elements such as formal and informal institutions, networks of state and non-state actors, and interactions between levels. Change in governance occurs through a combination of purposeful interaction and emergent properties resulting from self-organisation (Pahl-Wostl 2009; Stewart 2006). In her conceptualisation, learning is a stepwise process involving action, experimentation, and reflection, similar to the learning cycle. However, the ability of governance systems to deal with uncertainty and surprise through learning is often impeded by path dependencies and rule-bound management regimes, which give apparent stability to the system (Pahl-Wostl 2009).

Learning is often discussed as being a fundamental goal or process in the adaptive management of SES. Many discussions of adaptive management suggest that investing in social learning can transform complex problems. Nevertheless, whether and how social learning can be fostered in adaptive management is often addressed only vaguely (Armitage et al. 2008). Factors known to affect transformation in this aspect of SES include: the history of the situation such as path dependencies; institutional structures and policies; stakeholding and processes to facilitate learning. For example, Pahl-Wostl et al. (2012) demonstrate that in water governance, governance systems that are polycentric and well-integrated both vertically and horizontally, are both more conducive to learning, and more effective than systems with greater power differentials or fewer connections. In this book, we are interested in the potential of social learning to transform understanding of systems more broadly across the social ecological divide.

The literature on learning suggests a number of individual and group conditions that encourage double loop learning in individuals and allow it to be enhanced in group settings. Individual factors begin with existing frameworks or knowledge which mean new information is of interest, for example it is helpful to have mental models of SES as a starting point for new learning (Fazey et al. 2005). Learning requires new information, or variations in practice or experience that do not fit easily within existing frameworks, and which call them into question (Fazey et al. 2005; Mezirow 1997). Finally, individuals need dispositions for good thinking such as adventurousness, intellectual curiosity, a tendency to seek clarification and look for reasons, and to be reflective and meta-cognitive (aware of one's mental processes) (Fazey et al. 2005). Within groups, learning can be fostered by having a combination of diversity among members and trust among them (Curseu and Schruijer 2010), a relatively equal distribution of power among participants (Argyris 1976), a sense of interdependency, and a shared stake in solving a problem (Leeuwis 2004). Participants also need to have persuasive skills and enough time and space for exchange of ideas and debate in the group (Argyris 1976). The conditions for individual, social, and experiential learning provide us with a starting point for designing processes for critically examining our assumptions, and learning across the social ecological.

While the literature on learning outlined above highlights the potential for critical reflection in SES, it also suggests a need for more attention to where and how experiential and social learning occurs in the multiple practices of people involved with an SES. This would help us to move beyond the technically rational emphasis on disciplinary knowledge that is in much SES literature and is reflected in our initial description of *Tarerer*/Kelly Swamp in Chapter 2. Instead, we argue for an epistemology of practice, in which what we see and know is shaped by our practice, by what we do in the everyday, and to an extent by our thinking (reflection) as we are engaged in these actions (Schön 1991). Such an epistemology is consistent with the concepts of experiential and social learning outlined above, but also extends these two learning approaches this by drawing our attention to how learning occurs.

In the spontaneous activity of professionals, academics, and people in general, is a tacit knowing made up of many small interactions with context, ongoing recognitions of situations, judgements, and skilful performances that do not need to be consciously linked to theories or procedures. For example, an ecologist skilled in bird monitoring moves through *Tarerer*/Kelly Swamp recognising species, intuiting and feeling how to interact with individual birds to avoid disturbing them, and recording in ways that don't interrupt attention to these surroundings. Similarly, a social scientist maintains a lively attention during a conversation, hearing and interpreting concepts, and making quick judgements about when to intervene with a question or change of topic. For example, an environmental psychologist may interpret the values of individuals from what they say, without consciously thinking to do this, whereas a sociologist might easily engage with how people collectively remember the past (i.e., social memory), and how that shapes the present, or how such remembering or forgetting connects to landscape. In similar ways, a farmer manages feed for cattle, a cyclist remains upright on a trail, and a singer performs for an audience. An epistemology of practice encompasses multiple practice knowledges in relation to a system like *Tarerer*/Kelly Swamp, not only scientific 'disciplines' or technical skills, and provides a starting point for integrating these, for example to understand changes to the SES.

An epistemology of practice echoes, in part, Haraway's (1988) framing of situated knowledges, wherein what we observe is shaped by the gaze we bring to a context (our understanding of what we see), and in turn the tools that we have for understanding it, but then extends this further, in that all knowledge is inseparable from practice. Based in observations of professionals at work, Cook and Wagenaar (2012) argue for the primacy of practice, of which knowledge and context are constituent parts or artefacts. They draw on Bourdieu (1977) to conceptualise practice as an ongoing stream of activity, and expand on Polyani's identification of the importance of tacit knowledge, including experience, in addition to explicit knowledge, and on Dewey's (1988) understanding of 'inquiry' as active engagement with the world extending what one knows. Consistent with Pahl-Wostl (2009), the context that affords practice consists of dynamic interconnected networks as well as the taken-for-granted rules and procedures that make up institutions. Cook and Wagenaar (2012) define three key concepts. There is 'actionable understanding', a shared understanding developed of a situation to enable mutually acceptable action; and 'ongoing business', a flow of collective actions and experience within a context that affords this, often involving undocumented rules, procedures, and routines. Last there is 'the eternally unfolding present', in which knowledge is evoked within human consciousness through active engagement with the world, afforded by the context and briefly discussed earlier in relation to time. Cook and Wagenaar (2012) do not explicitly refer to 'learning' as a discrete process, however it is clear that knowledge develops through practice, a continual learning by doing. Ison et al. (2013, p. 35) describe a similar conceptualisation of knowledge in SES as an 'emergent relational dynamic' related to practice.

The ongoing business of the everyday does not require deliberate and conscious thinking about rules and procedures, however, there are times when people think about what they are doing, while they are doing it. Often this occurs when they are presented with something troubling, interesting or unusual that interrupts the usual flow of activity and causes them to reflect. They may ask themselves what features of the situation seem surprising or different, or what it is that they have just done that led to a positive or negative outcome, or whether a different approach might achieve a better outcome. This process of reflection in action is central to dealing with situations of uncertainty (Schön 1991), which are characteristics of complex systems. West et al. (2019) found taking an epistemology of practice approach was useful because it helped critical reflection on assumptions, guesses, and tacit understandings that underpin management in complexity. It exposed the stochastic rather than incremental nature of learning that is contingent on constellations of doing (p. 40). Further, it empowered scientists and managers to embrace contingencies as an intrinsic aspect of charting a course within complexity; and kept situations open that may foster creative new forms of practice. The epistemology of practice also has the potential to draw attention to those moments of dissonance that occur while engaging in research or management of SES that may be hidden during the integration of the rational and empirical that are normalised within conventional accounts (West et al. 2019).

Within research practice, action research (Reason and Bradbury 2008) give primacy to critical reflection on everyday practice. The development of pragmatism (see Dewey 1929, 1954) contributes to the centring of experience, the idea of starting from the bottom-up to communicate what is going on, and in communicating your ideas, to recognise multiple world views. Pragmatism recognises many of the same attributes to communication that we consider important in describing relationships within an SES. These include interconnectedness, uncertainty and a determination to overcome what Dewey considered the false emphasis on binaries (like mind/body) (Dewey 1929), as well as the gaze we bring to a context (our understanding of what we see), and in turn the tools that we have for understanding it.

Overall, our approach in this book is centred in an epistemology of practice and the critical reflection through which changes in practice emerge. By practice we mean any kind of ongoing everyday activity that researchers and others do within a system. In an epistemology of practice, knowledge is understood to be shaped by such actions and by our thinking (our reflections) while we do things (Schön 1991). Rather than something discrete, separate and applied, knowledge is a constituent of practice (Cook and Wagenaar 2012). Adopting an epistemology of practice in SES research means focussing on what researchers and others do within an SES, a stance that can make different knowledges (practices) readily accessible from multiple perspectives. Given our focus on SES, we are particularly interested in practices that are contributing to, elucidating or mitigating complex issues related to sustainability. To the focus on practice we add the insights from studies on learning within SES, which emphasise the need for higher level learning (double and triple loop learning) to open up our frameworks, values, and rules for critical analysis and change. In SES, social learning is crucial to integrating multiple practices, while dealing with uncertainty. We view social learning within an SES as a self-conscious process of articulation and recognition of a problem or situation that people are assembling together by demonstrating their practices or discussing what they 'do'. Through interaction, frames of reference and assumptions are opened to exploration and then to discussion about how our various actions might be different. Observation and reflection on action within the system sets up feedback loops, in which the consequences of different actions are laid out, and the most reasonable actions identified. Thus, learning occurs within a continuous cycle of practice, with ongoing learning an outcome of our practices. As systems are dynamic, the situation will necessarily be different in consequent actions, setting up the need for new social learning.

In the next chapter, we consider in more depth how social learning can be situated within cycles of practice. While learning is regarded as integral to adaptive management and adaptive governance (Armitage et al. 2008; Pahl-Wostl 2009), little attention has been given to how learning occurs. We explore questions about how learning interacts with the interruptions or shocks that happen within the ongoing flow of practice in an SES, and how within such cycles, we can intentionally create deliberative spaces that interrogate multiple understandings of SES and enable social learning. We frame the need for a deliberative space that fosters experiential social learning for all those involved in SES, and across multiple levels from individual to system processes, as central to reshaping methods and process in social ecological thinking.

References

Allison, H., and R. Hobbs. 2004. Resilience, Adaptive Capacity, and the "Lock-in Trap" of the Western Australian Agricultural Region. *Ecology and Society* 9 (1): 3.

Anderies, J.M., P. Ryan, and B. Walker. 2006. Loss of Resilience, Crisis, and Institutional Change: Lessons from an Intensive Agricultural System in Southeastern Australia. *Ecosystems* 9 (6): 865–878.

Anderson, B., M. Keanes, C. McFarlane, and D. Swanton. 2012. On Assemblages and Geography. *Dialogues in Human Geography* 2 (2): 171–189.

Anderson, B., and C. McFarlane. 2011. Assemblage and Geography. *Area* 43 (2): 124–127.

Angelstam, P., M. Elbakidze, R. Axelsson, N.E. Koch, T.I. Tyupenko, A.N. Mariev, and L. Myhrman. 2013. Knowledge Production and Learning for Sustainable Landscapes: Forewords by the Researchers and Stakeholders. *AMBIO* 42 (2): 111–115.

Argyris, C. 1976. Single-Loop and Double-Loop Models in Research on Decision-Making. *Administrative Science Quarterly* 21: 363–377.

Armitage, D., M. Marschke, and R. Plummer. 2008. Adaptive Co-management and the Paradox of Learning. *Global Environmental Change* 18: 86–98.

Aton, A. 2017. Earth Almost Certain to Warm by 2 Degrees Celsius. *Climate Wire*. https://www.scientificamerican.com/article/earth-almost-certain-to-warm-by-2-degrees-celsius/.

Ayre, M., and R. Nettle. 2015. Doing Integration in Catchment Management Research: Insights into a Dynamic Learning Process. *Environmental Science and Policy* 47: 18–31.

Bandura, A. 1971. *Social Learning Theory*. New York: General Learning Press.

Begon, M., J.L. Harper, and C.R. Townsend. 1990. *Ecology: Individuals, Populations and Communities*, 2nd ed. Cambridge, MA: Blackwell Scientific.

Beilin, R., and H. Bender. 2011. Interruption, Interrogation, Integration and Interaction as Process: How PNS Informs Interdisciplinary Curriculum Design. *Futures* 43: 158–165.

Berkes, F., J. Colding, and C. Folke (eds.). 2003. *Navigating Social-Ecological Systems: Building Resilience for Complexity and Change*. Cambridge, UK: Cambridge University Press.

Berkes, F., and C. Folke (eds.). 1998. *Linking Social and Ecological Systems: Management Practices and Social Mechanisms for Building Resilience*. Cambridge, UK: Cambridge University Press.

Berkes, F., and D. Jolly. 2001. Adapting to Climate Change: Social-Ecological Resilience in a Canadian Western Arctic Community. *Conservation Ecology* 5 (2): 18.

Binder, C.R., J. Hinkel, P.W.G. Bots, and C. Pahl-Wostl. 2013. Comparison of Frameworks for Analyzing Social-Ecological Systems. *Ecology and Society* 18 (4): 26.

Blaser, M. 2013. Ontological Conflicts and the Stories of Peoples in Spite of Europe. *Current Anthropology* 54 (5): 547–568.

Bourdieu, P. 1977. *Outline of Theory of Practice*. Cambridge, UK: Cambridge University Press.

Bracken, L.J., and E.A. Oughton. 2006. What Do You Mean? The Importance of Language in Developing Interdisciplinary Research. *Transactions of the Institute of British Geographers* 31 (3): 371–382.

Briassoulis, H. 2015. The Socio-Ecological Fit of Human Responses to Environmental Degradation: An Integrated Assessment Methodology. *Environmental Management* 56 (6): 1448–1466.

Briassoulis, H. 2017. Response Assemblages and Their Socioecological Fit: Conceptualizing Human Responses to Environmental Degradation. *Dialogues in Human Geography* 7 (2): 166–185.

Capra, F., and P.L. Luisi. 2014. *The Systems View of Life: A Unifying Vision.* Cambridge: Cambridge University Press.

Chaudhary, S., A. McGregor, D. Houston, and N. Chettri. 2015. The Evolution of Ecosystem Services: A Time Series and Discourse-Centered Analysis. *Environmental Science and Policy* 54: 25–34.

Climate Reality Project. 2017. Why People Ignore the Science Behind the Climate Crisis (And What You Can Do). https://www.climaterealityproject. org/blog/why-people-ignore-science-behind-climate-crisis-and-what-you-can-do.

Collard, R.-C., L.M. Harris, N. Heynen, and L. Mehta. 2018. The Antinomies of Nature and Space. *Environment and Planning E: Nature and Space* 1 (1–2): 3–24.

Cook, S.D.M., and H. Wagenaar. 2012. Navigating the Eternally Unfolding Present: Toward an Epistemology of Practice. *The American Review of Public Administration* 42 (1): 3–38.

Cooke, B., S. West, and W.J. Boonstra. 2016. Dwelling in the Biosphere: Exploring an Embodied Human-Environment Connection in Resilience Thinking. *Sustainability Science* 11: 831–843.

Cumming, G. 2014. Theoretical Frameworks for the Analysis of Social-Ecological Systems. In *Social-Ecological Systems in Transition*, ed. S. Sakai and C. Umetsu. Otsu, Japan: Springer.

Curseu, P., and S. Schruijer. 2010. Does Conflict Shatter Trust or Does Trust Obliterate Conflict? Revising the Relationships Between Team Diversity, Conflict and Trust. *Group Dynamics: Theory, Research and Practice* 14: 66–79.

Daly, M. 1978. *Gyn/Ecology: The Metaethics of Radical Feminism.* Boston: Beacon Press.

Datta, R. 2015. A Relational Theoretical Framework and Meanings of Land, Nature, and Sustainability for Research with Indigenous Communities. *Local Environment* 20 (1): 102–113.

DeLanda, M. 2006. *A New Philosophy of Society: Assemblage Theory and Social Complexity.* London: Continuum.

Dewey, J. 1929. *Experience and Nature.* New York, NY: W. W. Norton.

Dewey, J. 1954. *The Public and Its Problems.* Denver, CO: Alan Swallow (Originally published in 1927).

Dewey, J. 1988. *Reconstruction in Philosophy: Middle Works 1899–1924*, vol. 12. Carbondale, IL: Southern Illinois University Press (Original work published 1920).

Dixon, T.H. 2017. Curbing Climate Change: Why It's so Hard to Act in Time. *The Conversation.* http://theconversation.com/curbing-climate-change-why-its-so-hard-to-act-in-time-80117.

Dorward, A.R. 2014. Livelisystems: A Conceptual Framework Integrating Social, Ecosystem, Development, and Evolutionary Theory. *Ecology and Society* 19 (2): 44.

Dunn, G., R.R. Brown, J.J. Bos, and K. Bakker. 2017. Standing on the Shoulders of Giants: Understanding Changes in Urban Water Practice Through the Lens of Complexity Science. *Urban Water Journal* 14 (7): 758–767.

Durkheim, E. 1982. *The Rules of Sociological Method*, ed. S. Lukes. London, UK: The Free Press.

Engel, P.G.H., and M. Saloman. 1997. *Facilitating Innovation for Development: A RAAKS Resource Box*. Amsterdam: KIT Pub.

Environment Protection Authority (EPA). 2011. *How Will Climate Change Affect Victorian Estuaries?* Melbourne, VIC: Environment Protection Authority.

Fabinyi, M., L. Evans, and S.J. Foale. 2014. Social-Ecological Systems, Social Diversity, and Power: Insights from Anthropology and Political Ecology. *Ecology and Society* 19 (4): 28–40.

Farms, Rivers and Markets Project. 2012. *Farms, Rivers and Markets: Overview Report*. VIC. https://industry.eng.unimelb.edu.au/__data/assets/pdf_file/0007/2845609/farms-rivers-and-markets.pdf. Accessed 8 Oct 2018.

Fazey, I., J.A. Fazey, and D.M.A. Fazey. 2005. Learning More Effectively from Experience. *Ecology and Society* 10 (2): 4.

Ferreira, F.S. 2017. Critical Sustainability Studies: A Holistic and Visionary Conception of Socio-Ecological Conscientization. *Journal of Sustainability Education* 13: 1–22.

Fine, G.A., and T. McDonnell. 2007. Erasing the Brown Scare: Referential Afterlife and the Power of Memory Templates. *Social Problems* 54 (2): 170–187.

Freire, P. 1970. *Pedagogy of the Oppressed*. New York, USA: The Continuum International Publishing Group Inc.

Friis, C., and J.O. Nielsen. 2017. On the System: Boundary Choices, Implications, and Solutions in Telecoupling Land Use Change Research. *Sustainability* 9 (974): 1–20.

Gallopín, G.C. 1991. Human Dimensions of Global Change: Linking the Global and the Local Processes. *Global Environmental Change* XLIII (4): 3–17.

Gallopín, G.C., S. Funtowicz, M. O'Connor, and J. Ravetz. 2001. Science for the Twenty-First Century: From Social Contract to the Scientific Core. *International Social Science Journal* 53 (168): 219–229.

Gunderson, L., and C. Holling. 2002. *Panarchy: Understanding Transformations in Human and Natural Systems*. Washington, DC, USA: Island Press.

Haidt, J., and C. Joseph. 2004. Intuitive Ethics: How Innately Prepared Intuitions Generate Culturally Variable Virtues. *Daedalus* 133 (Fall): 55–67.

Haraway, D. 1988. Situated Knowledges: The Science Question in Feminism and the Privilege of Partial Perspective. *Feminist Studies* 14 (3): 575–599.

Haraway, D.J. 2008. *When Species Meet*. Minneapolis and London: University of Minnesota Press.

Hatt, K. 2013. Social Attractors: A Proposal to Enhance "Resilience Thinking" About the Social. *Society & Natural Resources* 26 (1): 30–43.

Head, L. 2016. *Hope and Grief in the Anthropocene*. London: Routledge.

Holling, C.S., L.H. Gunderson, and G.D. Peterson. 2002. Sustainability and panarchies. In *Panarchy: Understanding Transformations in Human and Natural Systems*, ed. L.H. Gunderson and C.S. Holling, 63–102. Washington, DC, USA: Island Press.

Ison, R., C. Blackmore, and B.L. Iaquinto. 2013. Towards Systemic and Adaptive Governance: Exploring the Revealing and Concealing Aspects of Contemporary Social-Learning Metaphors. *Ecological Economics* 87: 34–42.

Kearney, A.R. 1994. Understanding Global Change: A Cognitive Perspective on Communicating Through Stories. *Climate Change* 27: 419–441.

Keightley, E., and M. Pickering. 2012. *The Mnemonic Imagination: Remembering as Creative Practice*. Basingstoke, UK: Palgrave Macmillan.

Kharazzi, A., B.D. Fath, and H. Katzmair. 2016. Advancing Empirical Approaches to the Concept of Resilience: A Critical Examination of Panarchy. *Ecological Information, and Statistical Evidence Sustainability* 8 (9): 935.

Kolb, D. 1984. *Experiential Learning: Experience as the Source of Learning and Development*. Englewood Cliffs, NJ: Prentice Hall.

Krishnan, A. 2009. What Are Academic Disciplines? Some Observations on the Disciplinarity vs. Interdisciplinarity Debate. ESRC National Centre for Research Methods. University of Southhampton. Retrieved from http://www.forschungsnetzwerk.at/downloadpub/what_are_academic_disciplines2009.pdf.

Law, J. 1999. After ANT: Complexity, Naming and Topology. *The Sociological Review* 47 (1): 1–14.

Lazarus, E.D. 2017. Toward a Global Classification of Coastal Anthromes. *Land* 6 (13): 1–27.

Leduc Browne, P. 2018. Reification and Passivity in the Face of Climate Change. *European Journal of Social Theory* 21 (4): 435–452.

Leeuwis, C. 2004. *Communication for Rural Innovation*. Oxford, UK: Blackwell Science.

Liu, J. 2017. Integration Across a Metacoupled World. *Ecology and Society* 22 (4): 29.

Liu, J., T. Dietz, S.R. Carpenter, M. Alberti, C. Folke, E. Moran, A.N. Pell, P. Deadman, T. Kratz, J. Lubchenco, E. Ostrom, Z. Ouyang, W. Provencher, C.L. Redman, S.H. Schneider, and W.W. Taylor. 2007. Complexity of Coupled Human and Natural Systems. *Science* 317 (5844): 1513–1516.

Luthe, T. 2017. Success in Transdisciplinary Sustainability Research. *Sustainability (Switzerland)* 9 (1): 71.

Lyon, C., and J.R. Parkins. 2013. Toward a Social Theory of Resilience: Social Systems, Cultural Systems, and Collective Action in Transitioning Forest-Based Communities. *Rural Sociology* 78 (4): 528–549.

Meadows, D.H. 1982. Sustaining Tropical Forest Resources: A Systems Approach. Unpublished Manuscript Obtained from Jennifer Robinson. Resource Policy Center, Dartmouth College, Hanover, NH, USA.

Mezirow, J. 1997. Transformative Learning: Theory to Practice. *New Directions for Adult and Continuing Education* 74: 5–12.

Mezirow, J. 1998. On Critical Reflection. *Adult Education Quarterly* 48 (3): 185.

Mol, A. 1999. *The Body Multiple: Artherosclerosis in Practice.* Durham, NC: Duke University Press.

Moore, J. 2015. *Capitalism in the Web of Life: Ecology and the Accumulation of Capital.* London: Verso.

Odum, E. 1975. *Ecology: The Link Between the Natural and the Social Sciences,* 2nd ed. New York: Holt, Rinehart and Winston.

O'Malley, M. 2008. "Everything is Everywhere: But the Environment Selects": Ubiquitous Distribution and Ecological Determinism in Microbial Biogeography. *Studies in History and Philosophy of Biological and Biomedical Sciences* 39 (2008): 314–325.

Ostrom, E. 2009. A General Framework for Analyzing Sustainability of Social-Ecological Systems. *Science* 325 (5939): 419–422.

Pahl-Wostl, C. 2009. A Conceptual Framework for Analysing Adaptive Capacity and Multi-level Learning Processes in Resource Governance Regimes. *Global Environmental Change* 19: 354–365.

Pahl-Wostl, C., L. Lebel, C. Knieper, and E. Nikitina. 2012. From Applying Panaceas to Mastering Complexity: Toward Adaptive Water Governance in River Basins. *Environmental Science and Policy* 23: 24–34.

Phoenix, C., N.J. Osborne, C. Redshaw, R. Moran, W. Stahl-Timmins, M.H. Depledge, L.E. Fleming, and B.W. Wheeler. 2013. Paradigmatic Approaches to Studying Environment and Human Health: (Forgotten) Implications for Interdisciplinary Research. *Environmental Science and Policy* 25: 218–228.

Popper, K.R. 1963. *Conjectures and Refutations: The Growth of Scientific Knowledge.* New York: Routledge.

Prigogine, I. 1989. Thermodynamics and Cosmology. *International Journal of Theoretical Physics* 28: 927.

Rawluk, A., and A. Curtis. 2016. Reconciling Contradictory Narratives of Landscape Change Using the Adaptive Cycle: A Case Study from Southeastern Australia. *Ecology and Society* 21 (1): 17.

Reason, P., and H. Bradbury (eds.). 2008. *Sage Handbook of Action Research: Participative Inquiry and Practice*, 2nd ed. London: Sage.

Redman, C.L., and A.P. Kinzig. 2003. Resilience of Past Landscapes: Resilience Theory, Society, and the *longue durée*. *Conservation Ecology* 7 (1): 14.

Rockstrom, J., W. Steffen, K. Noone, A. Persson, F.S. Chapin III, E. Lambin, T.M. Lenton, M. Scheffer, C. Folke, H. Schellnhuber, B. Nykvist, C.A. De Wit, T. Hughes, S. van der Leeuw, H. Rodhe, S. Sorlin, P.K. Snyder, R. Costanza, U. Svedin, M. Falkenmark, L. Karlberg, R.W. Corell, V.J. Fabry, J. Hansen, B. Walker, D. Liverman, K. Richardson, P. Crutzen, and J. Foley. 2009. Planetary Boundaries: Exploring the Safe Operating Space for Humanity. *Ecology and Society* 14 (2): 32.

Saver, J.L., and A.R. Damasio. 1991. Preserved Access and Processing of Social Knowledge in a Patient with Acquired Sociopathy Due to Ventromedial Frontal Damage. *Neuropsychologica* 29 (12): 1241–1249.

Scholz, R.W. 2011. *Environmental Literacy in Science and Society: From Knowledge to Decisions*. Cambridge, UK: Cambridge University Press.

Schön, D.A. 1991. *The Reflective Practitioner: How Professionals Think in Action*. London, UK: Basic Books Inc.

Schusler, T., D. Decker, and M. Pfeffer. 2003. Social Learning for Collaborative Natural Resource Management. *Society and Natural Resources* 15: 309–326.

Sinclair, K., A. Curtis, E. Mendham, and M. Mitchell. 2014. Can Resilience Thinking Provide Useful Insights for Those Examining Efforts to Transform Contemporary Agriculture? *Agriculture and Human Values* 31: 371–384.

Smith, A., and A. Stirling. 2010. The Politics of Social-Ecological Resilience and Sustainable Socio-technical Transitions. *Ecology and Society* 15 (1): 11.

Spies, T.A., E.M. White, J.D. Kline, A.P. Fischer, A. Ager, J. Bailey, J. Bolte, J. Koch, E. Platt, C.S. Olsen, D. Jacobs, B. Shindler, M.M. Steen-Adams, and R. Hammer. 2014. Examining Fire-Prone Forest Landscapes as Coupled Human and Natural Systems. *Ecology and Society* 19 (3): 9–23.

Steg, L., A.E. Van den Berg, and J.I.M. De Groot. 2013. *Environmental Psychology: An Introduction*. Chichester, UK: BPS Blackwell.

Stewart, J. 2006. Value Conflict and Policy Change. *Review of Policy Research* 23 (1): 183–195.

Turner, B.L., R.E. Kasperson, P.A. Matson, J.J. McCarthy, R.W. Corell, L. Christensen, N. Eckley, J.X. Kasperson, A. Luers, M.L. Martello, C. Polsky, A. Pulsipher, and A. Schiller. 2003. A Framework for Vulnerability Analysis in Sustainability Science. *Proceedings of the National Academy of Sciences* 100 (14): 8074–8079.

Walker, B.H., N. Abel, J.M. Anderies, and P. Ryan. 2009. Resilience, Adaptability, and Transformability in the Goulburn-Broken Catchment, Australia. *Ecology and Society* 14 (1): 12.

Walker, B.H., L.H. Gunderson, A.P. Kinzig, C. Folke, S.R. Carpenter, and L. Shultz. 2006. A Handful of Heuristics and Some Propositions for Understanding Resilience in Social-Ecological Systems. *Ecology and Society* 11 (1): 13.

Walker, B.H., and D. Salt. 2006. *Resilience Thinking: Sustaining Ecosystems and People in a Changing World*. Washington, DC, USA: Island Press.

Walker, B.H., and D. Salt. 2012. *Resilience Practice: Building Capacity to Absorb Disturbance and Maintain Function*. Washington, DC, USA: Island Press.

Watts, V. 2013. Indigenous Place-Thought and Agency Amongst Humans and Non-humans (First Woman and Sky Woman Go on a European Tour!). *DIES: Decolonization, Indigeneity, Education and Society* 2 (1): 20–34.

Wernli, D., and F. Darbellay. 2016. Interdisciplinarity and the 21st Century Research-Intensive University: Pushing the Frontiers of Innovative Research. LERU. https://www.leru.org/files/Interdisciplinarity-and-the-21st-Century-Research-Intensive-University-Full-paper.pdf. Accessed 17 Oct 2018.

West, S., R. Beilin, and H. Wagenaar. 2019. Introducing a Practice Perspective on Monitoring for Adaptive Management. *People and Nature* 1: 387–405.

White, D., A. Rudy, and B. Gareau. 2017. *Environments, Natures and Social Theory: Towards a Critical Hybridity*. London: Palgrave Macmillan.

Wilson, E. O. 1998. *Consilience: The Unity of Knowledge* (No. 31). New York: Random House Digital.

CHAPTER 4

Adaptive Doing: Reimagining Social Ecological Practice

In this chapter, we present an approach for starting with practice in social ecological system (SES) research and disrupting the assumptions discussed in Chapter 3. A practice-oriented approach is not new. It has been argued for by others in different fields (e.g., Galafassi et al. 2017; Reason and Bradbury 2008) but it does not appear to be a common methodological approach in SES work at this time. As described in Chapter 3, practice and knowledge are irreducible in an epistemology of practice (Cook and Wagenaar 2012). What we do shapes what we come to know and the reverse. Taking a practice-oriented approach benefits SES thinking as an accessible entry-point to change how we understand and act in SES research. Different disciplinary knowledge with incommensurable philosophical underpinnings leads to challenging discussions. Instead of focusing on the philosophical basis of what is understood of SES research, by examining practice—*how and what individuals do*—we can allow different people to access understanding or knowing through describing what we do. A practice-oriented approach has the potential to overcome the disciplinary discourses that often divide researchers.

Putting practice first encourages alternative ways of seeing, knowing, and doing because it assists researchers to critically reflect on their assumptions and actions, which may be unconscious or unquestioned because they are normative for a discipline or culture, ultimately challenging the status quo. Maintaining this status quo, whether as an individual or in an organisational sense, requires a lot of work, as policy

© The Author(s) 2020
A. Rawluk et al., *Practices in Social Ecological Research*,
https://doi.org/10.1007/978-3-030-31189-6_4

scientists (Atkinson 2011) and critical theorists have argued (Foucault 1972). Assuming that language or observational status are intrinsically objective may just be contributing to the maintenance of oppressive situations, disguised as the status quo—as feminist linguistic scholars Burton (1982) and Daly (1978) argued so effectively.

Practices are products of the integration of rational, empirical, and experiential understanding (Cook and Wagenaar 2012). By focusing on practice in social ecological research, we draw attention to how we integrate these different aspects of knowing—rational, empirical, and experiential. We call this approach to focusing on practice in social ecological research 'adaptive doing'. Integration is crucial to practice and our engagement with SES. Turning our attention to 'adaptive doing' encourages researchers to take a transparent and holistic approach to practices and SES.

ADAPTIVE DOING: A PROCESS FOR CHANGING AND INTEGRATING KNOWLEDGE AND PRACTICE

Adaptive doing describes the common process of critically reflecting on and modifying or adapting our practices in response to context. It involves the integration of rational, empirical, and experiential understanding (Cook and Wagenaar 2012) and a more deliberate engagement with other ways of knowing, including but not limited to multiple disciplines. Such an interdisciplinary endeavour requires that the integration of disciplines is articulated as fully and transparently as possible. To overcome the disciplinary boundaries and the potential for divisional binaries between the social and ecological discourses that often isolate researchers, we aim to expand the outcomes from social ecological research by focusing on what really happens in the messiness of research and management. We need a process that allows researchers and managers to:

- engage with an SES as irreducible, emergent, and not entirely knowable;
- identify and reflect on differences in assumptions of reality, knowledge, and practice;
- explore the moral and ethical implications of the research with a desire to overcome binaries that limit understanding; and

- collaboratively create a shared, working understanding of the social ecological that is dynamic and therefore invites ongoing engagement.

Adaptive doing is also a product of an experiential learning process, which, according to Kolb (1984), has four phases: concrete experience (doing); reflective observation (reviewing); abstract conceptualisation (concluding); and active experimentation (planning). Integrating our understanding of SES with the phased nature of experiential learning, we have represented the adaptive doing learning cycle in Fig. 4.1 and labelled adaptive doing phases from A through to D, which we detail a little later in the chapter. In the process of adaptive doing, we identify and detail the agora, a deliberate space for critical reflection and creating change in knowledge and practice.

The Agora: a Deliberate Space for Creating Change in Understanding

In adaptive doing, the agora is in intentionally invoked space to challenge assumptions and facilitate change. Underpinning the agora is attention to *consciencisation* and integration, and it is invoked during the phases of critical reflection (A–C) in adaptive doing (see Fig. 4.1).

The term is inspired by the 'Agora' of Athens (Fig. 4.2), which was a meeting place with multiple uses in ancient Greece, where citizens could come together to discuss all topics, including politics, culture or the economy. It was regarded as a crucial space for the participation of citizens (Desouza and Bhagwatwar 2014). The Agora of Ancient Greece was a place in which practices and process emerged to construct a daily and iterative engagement with society. The Agora was not a building or physical structure that confined that space in a rigid way (Camp and Mauzy 2009). Some suggest that the openness and lack of structure of the Agora led to its ongoing potency. It is this fluidity, dynamic positioning and repositioning, and emergent properties that we wish to reference in using the term agora. In our imagining of the agora, it is a physical place or location that is invoked as needed and is not fixed in space or time. For the authors, this took various forms of meeting rooms at the university, the writing retreats for this book, as well as an online platform. It is also a mental state—a readiness to be challenged and changed, to have your assumptions questioned, and to be open to alternative ways of seeing and doing. It comes with a level of discomfort associated with

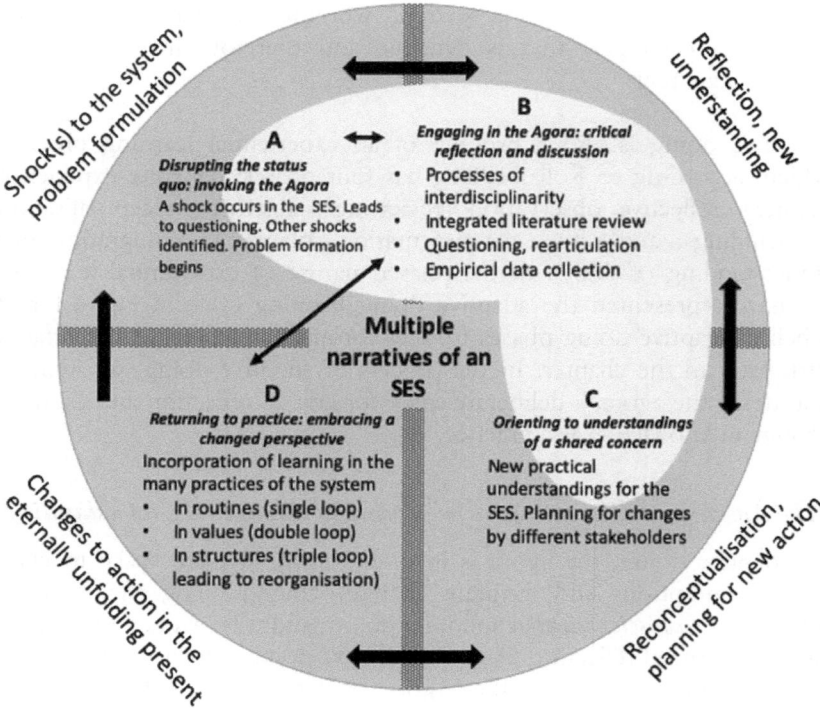

Fig. 4.1 Adaptive doing is a dynamic, four-phase, iterative process of learning and change. Within the adaptive doing process is the agora, which may be intentionally invoked and entered

a sense of uncertainty about the path forward, that is countered by a confidence in the collective to navigate this challenge.

In the mid-twentieth century, Hannah Arendt (1961) revitalised the idea of the agora to conceptualise a public space for political discussion. She envisioned it as a creative space that countered the barbaric and rigid politicisation of the totalitarianism of the earlier half century. Her writing indicates that the agora could be an ethically oriented space that enabled diverse perspectives and politics to come together for critical reflection. Politics would be at the heart of this space, with the intention of fostering equality (Arendt 1961). Though Arendt's (1998/1958) evocation of the agora is an idealised one, as authors, we nevertheless recognise

Fig. 4.2 The agora of Athens in Greece (Photo credit: Helena Bender, March 2019)

that her position on such a public space, as the site where individuals can affirm their potential for acting and thinking, is fundamental to changing what we do in the world. The agora is a space for understanding creativity and the inspiration for 'constructing' a commons that supports change. We draw on Arendt's ideas of the agora in describing our use— to provide the freedom for discussion and enacting of ideas, as well as representations of human and non-human engagement in the social ecological world. We also call on the power of its name as a symbol. We, as citizens and actors in a complex world, need to find the space to step into where there is room to reflect.

In recent years, the original Agora has inspired a way of describing digital places for discussion and learning. The connectivity and groups found in the Internet are often described as an agora, an unstructured and open space for science and society to meet. The Internet can be a place of meeting for scientific and local knowledge, where 'scientists and the public should meet on the "agora" to discuss important socio-scientific issues' (France et al. 2017, p. 323 citing Michael Gibbons). This 'digital agora' can be an emergent avenue for participation in a way that is accessible to many (Atifi and Marcoccia 2017), highlighting the multiple potential spaces for social learning and a space for leveraging democratic discussion, social change, and mobilisation that can lead to making and changing laws (Ranchordás 2017). In invoking the agora as a meeting place and a platform for fostering practice change, we build on its symbolic heritage as a space for discussion, learning, and as a stepping stone for action. In this way the agora turns participants outwards, providing context and perspectives that can inform changing practices in the everyday landscape.

Our conceptualisation of the agora reinforces the collective societal responsibility to discuss and arrive at various pathways for changing practice through listening to others. We turn here to the importance of the individual connecting with place through a deeper understanding of what it means to be in a place: in effect, knowing oneself differently, through an extension of experiencing that place. Japanese philosophy describes—the *basho*—and, though none of us as authors are philosophers, we offer our fledgling interpretation of its qualities as ones that strengthen the idea that, like non-human species, we are changed by how we interact with where we are in the world. While the idea of concientisation and the influences of the agora are predominantly political and outward looking, the *basho* is principally an internal reflection of a meditative state of immersion in the present moment.

Basho has many meanings that reflect different levels or scales of being or knowing. The *basho of being* refers to reflexive awareness, which assists us to gain knowing and insights from our everyday encounters that can bring us to question problematic subject/object distinctions (Chia 2003). At the scale of relationships, the *basho of relative nothingness* involves consideration of idealism and subjectivism because it challenges us to question whether 'experience' is something only individuals 'have'. Finally, at an infinite scale, the *basho of absolute nothingness* is the context in which all judgements are grounded; the 'place, the openness, the emptiness in which all particular occurrences are to be found, and yet is known only though their very occurrence' (Chia 2003, p. 971). Chia (2003) argues that it is instead a 'potentially fecund and pro-generative field of primordial knowing that inspires intervention, consciousness and understanding...where facts are encountered just as they are prior to our conceptual fabrications...there is no self, no thing, nothing separate or individual at all' (p. 971). In this philosophy, it is through the act of experiencing that the self is realised (Chia 2003). This metaphysical understanding of the *basho* goes beyond what we wish to imagine in describing the agora.

We adapt our use of *basho*, as others have done, to conceptualising it as a physical space or place, whose purpose is to provide a platform or meaning for the pursuit of or discussions about practice (Nonaka and Konno 1998); while fundamentally understanding that it offers an individual, an internal and profoundly meditative opportunity to experience change. In Chapter 5 one of the ecologists in our study of *Tarerer*/Kelly Swamp, brings a *basho*-like story to us in describing bird monitoring. In the authors' agora, that person's experience enriches our understanding of the connection between the individual and the *basho* as an individual place of change; as well as igniting much discussion about how such experience changes practice more generally.

Just as SES are not static in time and place, the agora is an emergent form. The enactment of the agora for social ecological research and the boundaries placed around it are as Law (1999), Mol (1999), and Blaser (2013) have written about in other contexts, effects and outcomes of our ideas and practices. As such, the agora is performed through the connectivity and dynamic interactions among human and non-human aspects of the SES, as Pickering (2010) described in his research on cybernetics. The knowable and inherently unknowable dimensions of an SES, and the knowable and unknowable gazes or understandings that each participant brings to the agora punctuate the ways in which we understand the SES. An SES is more than a sum of its parts; the interactions between the multiple known and unknown, knowable and unknowable aspects of an SES lead to emergent qualities, which 'cannot be accounted for by focusing on the human or non-human alone' (Pickering 2010, p. 196). Through discussion, participants in the agora may become more aware that all understandings of a social ecological context are incomplete and that this is a normative condition of acting in the present.

Engagement in the agora acknowledges the presence of power in all democratic spaces (Flyvbjerg 1998), where it is key to have as many voices and perspectives as possible in conversation (Mouffe 2000). This diversity can help avoid dominant ways of knowing and seeing from unconsciously pervading, even (and especially) in spaces of deliberation that are seen as neutral and building consensus; or as we argued earlier, reinforcing the status quo. Mouffe (2000) argues that such neutrality is impossible, and instead of trying to create spaces that are free of power, we must instead bring power to the centre of these discussions. The political space is not closed and exclusive, instead it is porous to new perspectives, ideas, and participants (Jones 2014). In social ecological contexts, bringing this diversity to discussions means implicating as many social, cultural, and scientific knowledges and practices as possible.

To summarise, our conceptualisation of the 'adaptive doing' agora is a symbolic and metaphorical space into which the narratives of those engaged with the messiness of SES thinking can come together to critically reflect, challenge, discuss, think, learn, and co-create working understandings that enable collaboration and investigation. Our intent in calling upon the agora is to go beyond discussion to facilitation, to action, and the fostering of a change in practice.

Navigating Adaptive Doing: Engaging Process in the Agora

The adaptive cycle is a commonly used heuristic in describing SES research, with its four phases of exploitation, conservation, collapse or reorganisation and release (see Walker and Salt 2006). It is often a 'shock' that triggers learning and practice change between the phases of the adaptive cycle. A shock may be a physical disturbance such as when repeat flooding and limited financial resources result in local graziers temporarily abandoning flood-prone fields as unsuitable for their business, as has occurred in *Tarerer*/Kelly Swamp. They may be social shocks, or the crossing of thresholds that are socially constructed and experienced (Christensen and Krogman 2012). In this framing, both physical and social shocks are seen as negative pressures or impacts that are 'caused by emergent hazards, vulnerability, or changes in risk tolerance' (Pelling and Manuel-Navarrete 2011), which can lead to collapse (Walker et al. 2004). In adaptive doing, we draw on a psychological understanding of disturbance, such that the trigger for engaging the agora occurs when an individual or community experiences something novel or unexpected in their ongoing experience of an environment, that need not be negative. We expand on this further in Phase A (see Fig. 4.1).

In adaptive doing, learning is associated with recognising something new or that something we have previously understood does not fit the current understanding. Adaptive doing is iterative and potentially ongoing, occurring through multiple learning cycles. The agora, however, need only be active during three phases of the adaptive doing experiential learning process (phases A through C, see Fig. 4.2). This learning can be in response to an interaction between humans. It may also be a response to an interaction between humans and non-humans, as described by one of our participants who changed their behaviour after observing shorebirds respond differently in varied contexts.

An individual or group may activate the agora. Although generally unquestioned, each participant comes to the agora with differing appreciations of reality and multiple ways of knowing that need to be articulated and discussed. Through a process of conscientisation, individuals become aware of these differences and may be open to changing practices as a consequence (Freire 1970). Remembering that the parts of the SES are irreducible, informs the respect that we each give to listening and understanding our different ways of grasping what we do and how such acts inform what we know. There will be challenges in comprehending each other. We do not intend the agora to reinforce existing or entrenched power dynamics, such as those based on assumptions that there is one way to do something. Therefore, there is a transparent identification and understanding of the multiple perceptions of what we do (in real time), how we make sense of our actions and how these contribute to our knowing. A critical outcome is the deliberative reimagining of the social ecological issues or problems, that led us to enter the agora in the first place. The aim is for transparency so that the different perceptions and dynamics present on entry, can be navigated openly, as Mouffe (2000) suggested with power, and then drawn on to form a transdisciplinary and/or interdisciplinary collaborative space.

Below we outline the processes and outcomes of each phase of the adaptive doing cycle and suggest reframing tools that might be used in these phases to overcome the binaries that commonly form in SES research.

Phase A—Disrupting the Status Quo: Invoking the Agora

Invoking the agora is brought about by novelty in, or a disruption of the status quo or way of understanding an SES. It may be inadvertent and similar to the sorts of shocks analysed with the adaptive cycle, like when neighbours in *Tarerer*/Kelly Swamp, contemplating rising flood waters, decided to dig a shallow drain to drain their paddocks. It may be noticing dissonance between participant narratives such as that led the authors to question the history and naming of the site or how dunes work as sponges that hold and release water. It may be a mismatch between physical observations and our internal representations of the world or our emotional responses. Making of space to discuss what is going on, to address issues of social ecological justice, or to generally understand issues in the SES with others, provides the catalyst for the agora to be invoked out of everyday practice.

In the section that follows, we draw upon literature about interactions between individual people's knowledge systems and their environment to consider in detail how entrenched experience and practices can be disrupted (Beilin and Bender 2011) and attention shifted to aspects of the SES that do not fit everyday expectations. Very often our conception of a context goes unquestioned until we face a shock or disturbance of some kind. We base our constructions on our knowledge and disciplinary cultures that shape what can be observed and how. In order to shift our understanding of and engagement in an SES, these perceptions can be explored in discussion, through describing what people do or suggest be done. Drawing on ideas in environmental psychology, human experience of the social ecological results from interaction between the external characteristics of the environment that people observe and their internal representations of the world, their mental maps or knowledge. Individuals' knowledge structures are built up through their multitude of past experiences and include two types of knowledge, characterised by typicality and novelty (Purcell 1992). Disruption in the first phase of adaptive doing involves interactions between these two types of knowledge and the physical and/or social environment.

The first type of knowledge is about what is 'typical' for an environment, it is based on regular, repeated experiences with similar environments and is represented internally as mental images of an expected or 'typical' environment. An example is a mental image of a rural landscape that is typical in the mind of the individual. The second type of knowledge is the understanding of how environments differ from the expected or typical. In contrast to representations of the typical, this consists of many representations of individual instances, as well as our knowledge of how variations can occur that are atypical. For example, this might include mental images of different landscapes and how they differ from the typical landscape. The two types of knowledge are richly interconnected, so that if one type is activated, this will make the other available (Purcell 1992). Among groups of people who interact, for example through shared interest or expertise, or in relation to a place, these knowledge structures are socially constructed, and continually re-constructed, through their everyday interactions. Importantly, diverse social representations of the same object may form within different social groups (Anderson et al. 2017). The interaction between these internal representations and changes to the external environment helps explain how everyday experience in the SES may be interrupted, catalysing a need to discuss these ideas in the agora.

In our interdisciplinary and transdisciplinary applied research practice, we have come to recognise the centrality of narratives as a common way in which very different participants can engage with issues. Narratives brought by participants provide rich descriptions of relationships, activities, and purposes for understanding an SES. Scientists and farmers alike have stories to tell about how things work and why that is the case. But as Wagenaar (2011, p. 210) points out, not everything is a story. To be a story it needs to be ordered in some way to seem a whole and as Czarniawska (2004) describes, there is attention to the plot. Furthermore, the storyline provides a way for the reader or listener to enter into the narrator's purpose. Multiple narratives of systems will potentially lead to different disruptions to what is known and seen, leading to the need for multiple ideas and the creation of the discussion space, the agora.

The dissonance that is felt, for example between narratives, can invoke a disruption of our status quo understanding of a context. Critical reflection occurs when there is a disruption to what seems typical and expected by something novel and unexpected in the ongoing experience of an environment. Where there is a fit between internal knowledge of what is typical and the external environment, there is a sense of familiarity and comfort, such that responses to typical landscapes are sometimes referred to as a 'warm glow of recognition' (Purcell 1992, p. 163). But where something novel is encountered, a mismatch or lack of fit occurs between the individual's external physical or/and social environment and the internal mental images that make up their knowledge.

The ongoing interactions between environment and knowledge are interrupted, leading to a physical response—the arousal of the autonomic nervous system—and subsequently, an emotional response. Where the mismatch is relatively small, the resulting feelings may be positive and curious, which may present us with something unexpected that arouses our interest and intrigues us with its novelty. If the mismatch is stronger, the experience is more intense, and the emotional response more unpredictable, with greater potential to be negative (Purcell 1992, 1993). The emotional response focuses our attention on whatever does not fit, which is followed rapidly by more cognitive efforts.

In the social ecological, this experience of mismatch can be the trigger that leads researchers to critically reflect on their assumptions, to identify problems and to invoke the agora. Experiences of encountering novel information are also essential to what work we undertake within the agora. As outlined in the previous chapter, learning is in the changes that occur when a person has novel experiences and develops a new understanding of their place in the world (Fazey et al. 2005), expanding their internal representations to be more inclusive, self-reflective, and integrative of experience (Mezirow 1997). For example, in *Tarerer*/Kelly Swamp, farmers with a focus on production may have a typical mental image of the Swamp landscape, particularly the floodplain, with less winter water, intending to make the Swamp and floodplains accessible to cattle and management vehicles. Consequently, they act to improve production values, optimising pasture access, even if early draining of the floodplain disrupts certain non-human species. As Anderson et al. (2017) argue, negative impacts for valued aspects of the SES can motivate a range of actions; including, we add, engagement in the agora, if only to renegotiate conceptions of what 'fits'.

Phase B—Engaging in the Agora: Critical Reflection and Discussion

The agora as an intentionally invoked and actioned space for engaging in critical reflection entails a commitment to leaving it recognising the potential and paths for change in one's knowledge or practice. Engaging in the agora is not simply a discussion; engaging in the agora is about entering with the knowledge that different understandings will emerge. Experiencing novelty may galvanise or encourage individuals or groups to participate in the agora. Ideally, participants in the agora would benefit from an openness to learning, a willingness to challenge assumptions, and an awareness of power relationships and processes for navigating these challenges. But equally they may be unexpectedly changed by the experience of being there and what they learn. Participants initially describe their skills and expertise, and then consider how those shape what is and is not seen in an SES (Haraway 1988; Cook and Wagenaar 2012). We recognise the importance of developing practice that allows us to make our assumptions about knowledge, reality, and previous practice transparent. Those with experience in the agora, enter in full awareness of the disciplinary lenses and assumptions in their 'toolboxes' and the intention to participate, listen, exchange ideas, and learn.

Acknowledging that the agora is an inherently political space, raises a related consideration about the size of the group participating in the agora. While a large group may provide more resources and diverse insights to spark social learning, this can also add to the complexity and difficulty of discussions (Wernli and Darbellay 2016). Smaller groups can also have difficulties due to lack of diversity or resources. Either can increase the risk of losing trust among participants. Throughout these processes, an awareness of power is important, and attention to power dynamics among participants can foster productive interactions (Argyris 1976).

Once a group of collaborators is established, it is helpful to build a shared understanding of the ways the group will operate. This may include planning the processes by which the group will work, such as meetings, workshops, or other sites for integration of ideas and issues, while also leaving space and flexibility for social learning and for the process to emerge in unexpected directions (Ayre and Nettle 2015). Building a shared understanding of process may involve rulemaking among participants, for example agreeing to avoid potentially isolating disciplinary jargon (Ostrom 1990; Wernli and Darbellay 2016).

While multiple narratives can make us aware of disruptions (Phase A), in Phase B, narratives and narrative analysis can provide a way to ensure that the process in the agora is accessible and inclusive to all people, and not only those with a particular knowledge culture. Even the most artlessly told narrative is connected to the wider life of that person, their places and their experiences. Wagenaar says: 'we cannot see stories apart from everyday practical situations...the relationship is reciprocal if not dialectical...the one brings the other into being' (2011, p. 211). Narrative and narrative analysis have particular connotations and application in some social research (e.g. Czarniawska 2004) that can be drawn on to analyse the discussions if there is interest and capacity to do so. As each participant articulates their story or represents voices and others associated with a context, the overall appreciation of how these relationships are formed and the meanings associated with the site will change. Critically engaging with and reflecting on these multiple connections, eliciting the quality of relationships, articulating an understanding of multiple perspectives are all ways of exploring an SES, this messy entanglement, in the agora (Jones 2009). Knowledge in and of SES is an 'emergent relational dynamic' of multiple practices and social interactions (Ison et al. 2013, p. 35).

There are different tools and devices that can be drawn on in the agora that will shape the understanding of those present, and the development of new shared narratives. When academics are the dominant participants in an agora, an integrated literature review (Wernli and Darbellay 2016) can be seen as a device that is understood by all researchers, though they may see very different meanings in the contents (Wernli and Darbellay 2016), which then become part of the discussion. Local and expert knowledge, as narratives, can also inform participants' understanding of the context.

We return here to the description of and justification for critical reflection that we outline in Chapter 3. From the field of adult education, Mezirow (1998) describes how critical reflection involves a deliberate choice to engage in a discourse considering alternative understandings. Participants undergo a process of critical reflection that leads them to explore the often-unarticulated assumptions of reality, knowledge, and practice that we each hold, as a starting point to both changing ourselves and our world (Freire 1970).

Concepts or ideas can be drawn on as triggers for critical reflection. In Chapter 3, we identify particular ideas that are often unarticulated in the study of SES, such as time. We noted that changing timelines, and our conception of time, triggers very different outcomes in the social ecological realm. In other research we have undertaken, we note the value of having a 'social historian' within the agora, a designated participant who keeps a journal of the ideas so that the participants may return to them to elucidate triggers or note moments of change. Such a document could be confirmed, like minutes at the start of a meeting, and would also be subject to house-rules, in the sense that if it is used, then the participants must agree how it will be used and whether it will exist past that engagement in the agora. These ideas can be useful to trigger critical reflection among participants.

Re-examining a system from a different focal position can also provoke the questioning of an understanding and facilitate critical reflection. A system-level analysis examines the structure and function of the SES, the components of the system and how they relate. Considering how this scale is engaged with and understood across multiple participants can give rise to multiple formulations of a system and identify new connections. A relationship level could involve considering how different participants enact the assembling of elements creating and analysing relationships for exploration or understanding. Reflection on relationships can also consider the political, cultural, social, or disciplinary drivers that shape how people understand and assemble an SES. A local-level analysis can explore how individuals embody and enact 'know-how' (Bourdieu 1977) within an SES. Social ecological research has typically engaged a more objective viewpoint, and critically reflecting on practice as a personal and subjective experience in an SES can provide multiple lenses.

Questions that can assist with critical reflection include:

- How are the narratives of the SES similar and different between participants? How are they different to your own?
- What connections become evident in listening/reading the multiple narratives?
- Is there anything or anyone not yet included in the narratives?
- How would your understanding of the SES change if the boundaries were made larger/smaller in space or time?
- How does changing timelines alter the way you understand the SES?
- How do you expect the SES to be in 50 years? 100 years?

The agora provides the space and place in which we can examine and reassess what we do and why through our practice lens. Critical reflection provides the deliberative opportunities within this process to listen, observe, and analyse multiple understandings of the SES. Attending to structure, function, relationships, space, time, and boundaries are all ways of instigating and following narrative meaning towards heightened awareness and the possibilities of changing practices.

Phase C—Orienting to Understandings of a Shared Concern

While peoples' perspectives and understanding (narratives) may be different, the purpose of the two previous phases is to identify the different understandings around a shared concern. Phase C does not imply consensus, but may involve some shared understanding, and allows researchers to step back into their day-to-day practice.

There is not a single way for participants to create a shared understanding of the SES or to orient to understandings of a shared concern, but there are certain aspects that need to be considered in doing so. Interdisciplinary integration in the agora is a social process of integration among academic disciplines (Ayre and Nettle 2015). To the extent that others are involved (e.g., resource users, interest groups), engagement in the agora also has elements of collective action, such that insights are relevant from experience with co-management of environments by groups of people with different histories, knowledges, and instruments who work towards a common purpose (e.g., Borrini-Feyerabend et al. 2000; Leeuwis 2004; Ostrom 1990). Creating a shared understanding relies on integrated and social ecologically just (Collard et al. 2018) ways of representing the entirety of the system, human and non-human. Integration can use material elements or 'devices' such as models and texts, around which interdisciplinary meanings are built (Ayre and Nettle 2015). Devices can also be 'boundary objects' that are recognisable from different perspectives and provide a starting point for interdisciplinary conversations (Bracken and Oughton 2006). In research contexts, a range of devices can provide a focus for developing a shared understanding, including joint research questions, integrative literature reviews (Wernli and Darbellay 2016) and action research methodology, in which researchers play dual roles of working towards practice changes and reflecting on how these occur (Reason and Bradbury 2007). Other devices can facilitate collaborative decision-making. For example, when negotiating in relation to system outcomes, interim negotiating criteria (Trainor 2006; Leeuwis 2004) or principles that communicate what is valued in the system (Schirmer et al. 2016) can guide the process. Good process can help to maintain productive activity, but where issues arise, formal conflict management may be needed to facilitate integrative negotiation (Chamala and Mortiss 1990).

Social learning about the social ecological is associated with integrative negotiation. In contrast to compromises in which both parties give something up, integrative negotiation involves participants shifting their epistemological frames of reference to enable the creation of new shared ways of framing, and creative ways forward (Leeuwis 2004). For example, Ayre and Nettle (2015) describe how a group of researchers working together on a large water catchment system (The Murray–Darling Basin) identified a previously unanticipated need to examine trade-offs in both the environmental and agricultural consequences of different water sharing options (Ayre and Nettle 2015). In another, simpler example, productive integration may begin with identifying and exploring, through questioning, the different disciplinary meanings of commonly used terms, such as 'catchment'. Bracken and Oughton (2006) use the example of physical and social scientists having different understandings and perceived relevance of a 'catchment' (p. 378). The physical scientist described a catchment as the drainage basin to a river, and while the social scientists accepted this, they found more relevance in the social interactions of people with landscape and the team needed to form a more complex and shared understanding of catchment through 'clarifying, justifying and arguing' (p. 378). This kind of exploration of language is a powerful device for exploring disciplinary assumptions before arriving at new shared definitions that can be used later in the process (Bracken and Oughton 2006).

Scholars of interdisciplinarity stress that a careful exploration of assumptions should be part of any attempts at integration. MacMynowski (2007) writes that only after articulating and clarifying assumptions of reality, knowledge and practice, should participants begin the process of synthesising how the different pieces of the collective understanding fit together. Bracken and Oughton (2006) express this as a process of initially articulating, followed by rearticulation of a shared understanding of a context. McDonald et al. (2009) explore a wide range of dialogue-based methods with various origins, that can be applied to research integration. For integrating visions worldviews and values, they suggest methods such as appreciative inquiry and principled negotiation. In adaptive doing, the formation of a shared understanding occurs only after critical reflection among individuals about how they each understand the social ecological context, and then how they can collectively understand it.

Navigating interpersonal challenges and power dynamics can be essential for both exploring multiple understandings in Phase B and creating a shared understanding in Phase C. These might be encountered, for example, through differences in knowledge cultures between social and biophysical researchers and other participants (McMynowski 2007) or value differences that may be incommensurable (Trainor 2006). Productive engagement can be supported with leadership or facilitation that can manage interaction, communication, and teambuilding (Wernli and Darbellay 2016). Facilitators need characteristics such as credibility, skills related to the dynamics of learning and how to intervene in them, flexibility, and independence (Leeuwis 2004). Loosely structured networks are frequently more conducive to creativity and learning than rule-bound ones (Pahl-Wostl 2009) and facilitators must exercise good judgement about the timeframes and structures that enable interactions.

Three Reframing Tools for Fostering Critical Reflection

There are multiple ways that the disruption of the status quo and critical reflection can occur. Here we draw on three approaches from the social sciences that we position as reframing tools: the 4 Is reframing tool; assemblage; and the eternally unfolding present. Each of these reframing tools address different assumptions common in SES research, as outlined in Chapter 3, and utilise narratives to expose different aspects of an SES. The 4 Is reframing tool assists to investigate the multiple factors that may affect the structure and function of the system. Assemblage takes a relational view of the SES to redirect efforts on bridging the separation between the elements. The eternally unfolding present helps to explore the importance of everyday actions in relation to time to focus on the minutia to see its effects on the larger patterns of research and management practice. Each of these reframing tools enables those engaging in adaptive doing to observe a system from a different perspective and form a new understanding.

We use the **4 Is** reframing tool to challenge the assumptions of structure and function in an SES so that integration at the system scale can occur. Beilin and Bender (2011) initially developed the 4 Is method—interruption, interrogation, integration, and interaction—for critically engaging in interdisciplinary pedagogy with university students. The intention was to empower students so that they could take their learnings out into the broader world through practices that assisted with creating change (Beilin and Bender 2011). **Interruption** is of the dominant understanding of a system or position. **Interrogation** entails changing the scale or boundaries to understand the many connections within an SES in different ways. **Interaction** considers the relationships between the system and other systems at different spatial scales, and how they are managed. **Integration** requires an examination of how multiple perspectives on an SES, held by different people or at different scales, can coalesce around a shared understanding to create individual and collaborative responsibility towards practice change.

As a structured form of critical reflection, the 4 Is are informed by critical social theory as well as post-normal science, where facts are not certain and there are multiple interpretations of social ecological challenges. The 4 Is are generally undertaken at a formative stage and at a systems level, when considering how to reimagine or understand why there are multiple views about the possible SES and how these can be integrated into ongoing ways of operating in research and in practice with academic and non-academic others. Diverse knowledges from the extended peer community are valued, including those that are not considered 'academic' and arise from experience or are understood as 'common sense' (Newell et al. 2005; Funtowicz and Ravetz 1994). This inclusivity may be understood as embracing transdisciplinary (non-academic or lay experts) and interdisciplinary voices. We apply the 4 Is as one process for critical reflection to *Tarerer*/Kelly Swamp in Chapter 5.

Critically examining multiple understandings of an SES using the 4 Is reframing tool can include analysing structure and function. How non-humans are represented in the SES could be one structural consideration. Functional considerations in an SES might involve: how government is present in the system and how it acts; the broad drivers that shape this system (e.g., national and international markets); and what values are present in the landscape. These can be compared across multiple participants in the agora.

As a reframing tool, **assemblage** assists us to engage with assumptions of knowledge and ordering by focusing on relationships. An assemblage is a focused network of relationships. There are many different assemblages in an SES. For example, in *Tarerer*/Kelly Swamp two of these assemblages include migratory birds and farmers (see Chapter 5). To take an approach of critical reflection to these assemblages, one element is brought into focus and consideration is given to how it changes each of the relationships (Briassoulis 2015). Each narrative will generally express a single focus for the described relationships. In such, discussion of different understandings of an SES and critical reflection on how varying narratives differ can illuminate alternative ways of understanding a system.

The **eternally unfolding present** (EUP) helps to challenge assumptions of SES research by centring on the present moment to experience timelessness and the irreducibility of the social ecological. In the EUP, one can imagine and inhabit the immediate experience of an individual, stripped of theory. The idea of the eternally unfolding present (EUP) comes from a Japanese philosophical tradition as described by Cook and Wagenaar (2012) and used as an analytical method by West (2015) and West et al. (2019) for considering micro-level individual actions. An EUP analysis is underpinned by three main ideas that need to be articulated and critically reflected on: actionable understanding; ongoing business; eternally unfolding present. Identifying an **actionable understanding** involves describing how people define what they are doing, the aims of the project and their role within it, how they each conceptualise the situation, come to decisions, and then generate practice. **Ongoing business** reflects the way in which people conduct their everyday practices, or the ongoing way in which these contribute to the function of the SES. Ongoing business includes environmental processes, job roles, project procedures, routine actions, rules and requirements, skills, tools, understandings of practice, and disruptions to these practices. The third idea is one that also acknowledges the importance of time. In the **eternally unfolding present** participants critically engage with the context and place of practice, the knowledge that underpins practice (all the ways that they 'know' and 'act' to create an awareness, the consciencisation), and the expectation of timelessness.

These different reframing tools are not intended to be prescriptive or rigid, but offer a starting point for critically reflecting on assumptions common in SES research. Participants in the agora can draw on these or find other ways of critically reflecting on the system of interest. These reframing tools can be used individually or in combination. In the following chapter we demonstrate both individual and combined applications.

Phase D—Returning to Practice: Embracing a Changed Perspective

In identifying Phase D as a distinct phase in the cycle of adaptive doing, we recognise that much of daily life involves an ongoing flow of activity that is dynamic in the sense that it is made up of many small interactions which, combined with the background we bring to action, creates context; and, the ongoing recognitions of situations and judgements about what to do next. Knowledge is tacit but may become conscious in moments when we think about what we are doing (Schön 1991). After engagement in the agora, participants in Phase D return to their everyday practice with changed perspectives and processes. They may entertain different thoughts as they go about their normal activities, with moments of critical reflection interspersed with their doing, such as: *What new knowledge or understanding have I gained that changes how I go about my everyday life and/or work?* For example, a farmer may leave certain parts of a paddock unmown, either to implement an agreement made in Phase C, or in response to a passing thought that occurs during mowing: *given that conversation, what if I do this differently?* Other changes to practice in Phase D may emerge as a result of more complex governance changes made by government institutions following agreements in Phase C. As academics engaged in research who are participating in Phase D, this might look like changes to teaching practice, or to the design of research.

In this part of the cycle, participants may also reflect on the kind of learning process they have been involved in and the extent of change that has occurred. Participants in the agora may reflect on whether changed practices reveal single loop learning (changes in routines consistent with governing frameworks), or double loop learning (changes in frameworks, values, and beliefs) or even triple loop learning (changes to governance norms, rules, and protocols) (Armitage et al. 2008). They may also be aware of being engaged in multiple cycles with different timescales that intersect in the eternally unfolding present.

In this chapter, we have introduced the reader to adaptive doing and the agora as a part of reorienting social ecological research to focus on practice. We have described the four iterative and dynamic phases of adaptive doing and offer three reframing tools as suggestions for examining a different focal position of a system and the experience of practice from that position. These reframing tools are the 4 Is, assemblage, and the eternally unfolding present. Utilising these reframing tools can enable participants in the agora to see the system in a new way and allow them to return to their own practices, changed. In the final chapter, we apply adaptive doing and draw on the three reframing tools described above in *Tarerer*/Kelly Swamp.

REFERENCES

Anderson, N.M., R.M. Ford, and K.J.H. Williams. 2017. Contested Beliefs About Land-Use are Associated with Divergent Representations of a Rural Landscape as Place. *Landscape and Urban Planning* 167: 75–89.

Arendt, H. 1961. The Concept of History. In *Between Past and Future*. New York, NY: Viking Press.

Arendt, H. 1998/1958. *The Human Condition*, 2nd ed. Chicago and London: Chicago University Press.

Argyris, C. 1976. Single-Loop and Double-Loop Models in Research on Decision-Making. *Administrative Science Quarterly* 21: 363–377.

Armitage, D., M. Marschke, and R. Plummer. 2008. Adaptive Co-management and the Paradox of Learning. *Global Environmental Change* 18: 86–98.

Atifi, H., and M. Marcoccia. 2017. Exploring the Role of Viewers' Tweets in French TV Political Programs: Social TV as a New Agora? *Discourse, Context & Media* 19: 31–38.

Atkinson, M. 2011. Lindblom's Lament: Incrementalism and the Persistent Pull of the *status quo*. *Policy and Society* 30 (1): 9–18.

Ayre, M., and R. Nettle. 2015. Doing Integration in Catchment Management Research: Insights into a Dynamic Learning Process. *Environmental Science and Policy* 47: 18–31.

Beilin, R., and H. Bender. 2011. Interruption, Interrogation, Integration and Interaction as Process: How PNS Informs Interdisciplinary Curriculum Design. *Futures* 43: 158–165.

Blaser, M. 2013. Ontological Conflicts and the Stories of Peoples in Spite of Europe. *Current Anthropology* 54 (5): 547–568.

Borrini-Feyerabend, G., M. Taghifavar, J.C. Nguinguiri, and V. Ndangang. 2000. *Environmental Management Co-management of Natural Resources Organising, Negotiating and Learning-by-Doing*, Eschborn, IUCN, GTZ.

Bourdieu, P. 1977. *Outline of Theory of Practice*. Cambridge, UK: Cambridge University Press.

Bracken, L.J., and E.A. Oughton. 2006. What Do You Mean? The Importance of Language in Developing Interdisciplinary Research. *Transactions of the Institute of British Geographers* 31 (3): 371–382.

Briassoulis, H. 2015. The Socio-Ecological Fit of Human Responses to Environmental Degradation: An Integrated Assessment Methodology. *Environmental Management* 56 (6): 1448–1466.

Burton, D. 1982. Through Glass Darkly, Through Dark Glasses. In *Language and Literature*, ed. R. Carter. London: Allen and Unwin.

Camp, J.Mc.K., and C.A. Mauzy. 2009. *The Athenian Agora: New Perspectives on an Ancient Site*. Mainz am Rhein: Verlag Philipp von Zabern.

Chamala, C., and P. Mortiss. 1990. *Working Together for Landcare: Group Management Skills and Strategies*. Brisbane: Australian Academic Press.

Chia, R. 2003. From Knowledge-Creation to the Perfection of Action: Tao, Basho and Pure Experience as the Ultimate Ground of Knowing. *Human Relations* 56 (8): 953–981.

Christensen, L., and N. Krogman. 2012. Social Thresholds and Their Translation into Social-Ecological Management Practices. *Ecology and Society* 17 (1): 5.

Collard, R.-C., L.M. Harris, N. Heynen, and L. Mehta. 2018. The Antinomies of Nature and Space. *Environment and Planning E: Nature and Space* 1 (1–2): 3–24.

Cook, S.D.M., and H. Wagenaar. 2012. Navigating the Eternally Unfolding Present: Toward an Epistemology of Practice. *The American Review of Public Administration* 42 (1): 3–38.

Czarniawska, B. 2004. *Narratives in Social Science Research*. London: Sage.

Daly, M. 1978. *Gyn/Ecology: The Metaethics of Radical Feminism*. Boston: Beacon Press.

Desouza, K., and A. Bhagwatwar. 2014. Technology-Enabled Participatory Platforms for Civic Engagement: The Case of U.S. Cities. *Journal of Urban Technology* 21 (4): 25.

Fazey, I., J.A. Fazey, and D.M.A. Fazey. 2005. Learning More Effectively from Experience. *Ecology and Society* 10 (2): 4.

Flyvbjerg, B. 1998. *Rationality & Power: Democracy in Practice*. Chicago, USA: University of Chicago Press.

Foucault, M. 1994/1972. *Archaeology of Knowledge*. New York, USA: Tavistock.

France, B., S. Birdsall, and L. Simonneaux. 2017. Analysing the Multiplicity of Voices in the Agora: Using Actor-Network Theory to Unravel a Complex Issue. *International Journal of Science Education, Part B* 7 (4): 323–340.

Freire, P. 1970. *Pedagogy of the Oppressed*. New York: Herder and Herder.

Funtowicz, S., and J.R. Ravetz. 1994. The Worth of a Songbird: Ecological Economics as a Post-normal Science. *Ecological Economics* 10: 197–207.

Galafassi, D., T. Daw, L. Munyi, K. Brown, C. Barnaud, and I. Fazey. 2017. Learning About Social-Ecological Trade-Offs. *Ecology and Society* 22 (1): 2.

Haraway, D. 1988. Situated Knowledges: The Science Question in Feminism and the Privilege of Partial Perspective. *Feminist Studies* 14 (3): 575–599.

Ison, R., C. Blackmore, and B.L. Iaquinto. 2013. Towards Systemic and Adaptive Governance: Exploring the Revealing and Concealing Aspects of Contemporary Social-Learning Metaphors. *Ecological Economics* 87: 34–42.

Jones, O. 2009. After Nature: Entangled Worlds. In *A Companion to Environmental Geography*, ed. N. Castree et al. Oxford: Wiley-Blackwell.

Jones, M. 2014. Chantal Mouffe's Agonistic Project: Passions and Participation. *Parallax* 20 (2): 14–30.

Kolb, D. 1984. *Experiential Learning: Experience as the Source of Learning and Development*. Englewood Cliffs, NJ: Prentice Hall.

Law, J. 1999. After ANT: Complexity, Naming and Topology. *The Sociological Review* 47 (1): 1–14.

Leeuwis, C. 2004. *Communication for Rural Innovation*. Oxford, UK: Blackwell Science.

MacMynowski, D.P. 2007. Pausing at the Brink of Interdisciplinarity: Power and Knowledge at the Meeting of Social and Biophysical Science. *Ecology and Society* 12 (1): 20.

McDonald, D., G. Bammer, and P. Deane. 2009. *Research Integration Using Dialogue Methods*. Canberra: ANU Press.

Mezirow, J. 1997. Transformative Learning: Theory to Practice. *New Directions for Adult and Continuing Education* 74: 5–12.

Mezirow, J. 1998. On Critical Reflection. *Adult Education Quarterly* 48 (3): 185.

Mol, A. 1999. *The Body Multiple: Artherosclerosis in Practice*. Durham, NC: Duke University Press.

Mouffe, C. 2000. *The Democratic Paradox*. London and New York: Verso.

Newell, B., C. Crumley, N. Hassan, E. Lambin, C. Pahl-Wostl, A. Underdal, and R. Wasson. 2005. A Conceptual Template for Integrative Human-Environment Research. *Global Environmental Change* 15: 299–307.

Nonaka, I., and N. Konno. 1998. The Concept of 'Ba': Building a Foundation for Knowledge Creation. *California Management Review* 40 (Spring): 40–54.

Ostrom, E. 1990. *Governing the Commons: The Evolution of Institutions for Collective Action*. Cambridge, UK: Cambridge University Press.

Pahl-Wostl, C. 2009. A Conceptual Framework for Analysing Adaptive Capacity and Multi-level Learning Processes in Resource Governance Regimes. *Global Environmental Change* 19: 354–365.

Pelling, M., and D. Manuel-Navarrete. 2011. From Resilience to Transformation: The Adaptive Cycle in Two Mexican Urban Centers. *Ecology and Society* 16 (2): 11–22.

Pickering, A. 2010. *The Mangle of Practice: Time, Agency, and Science*. Chicago: University of Chicago Press.

Purcell, A.T. 1992. Abstract and Specific Physical Attributes and the Experience of Landscape. *Journal of Environmental Management* 34: 159–177.

Purcell, A.T. 1993. Relations between Preference and Typicality in the Experience of Paintings. *Leonardo* 26 (3): 235–241.

Ranchordás, S. 2017. Digital Agoras: Democratic Legitimacy, Online Participation and the Case of Uber-Petitions. *The Theory and Practice of Legislation* 5 (1): 31–54.

Reason, P., and A. Bradbury. 2007. *Handbook of Action Research*, 2nd ed. London, UK: Sage.

Reason, P., and H. Bradbury (eds.). 2008. *Sage Handbook of Action Research: Participative Inquiry and Practice*, 2nd ed. London: Sage.

Schirmer, J., M.L. Dare, and S.A. Ercan. 2016. Deliberative Democracy and the Tasmanian Forest Peace Process. *Australian Journal of Political Science* 51 (2): 288–307.

Schön, D.A. 1991. *The Reflective Practitioner: How Professionals Think in Action*. London, UK: Basic Books Inc.

Trainor, S.F. 2006. Realms of Value: Conflicting Natural Resource Values and Incommensurability. *Environmental Values* 15: 3–29.

Wagenaar, H. 2011. *Meaning in Action: Interpretation and Dialogue in Policy Analysis*. Armonk, NY and London, England: M.E. Sharpe.

Walker, B., C.S. Holling, S.R. Carpenter, and A. Kinzig. 2004. Resilience, Adaptability and Transformability in Social–Ecological Systems. *Ecology and Society* 9 (2): 5–14.

Walker, B., and D. Salt. 2006. *Resilience Thinking: Sustaining Ecosystems and People in a Changing World*. Washington, DC, USA: Island Press.

Wernli, D., and F. Darbellay. 2016. Interdisciplinarity and the 21st Century Research-Intensive University: Pushing the Frontiers of Innovative Research. LERU. https://www.leru.org/files/Interdisciplinarity-and-the-21st-Century-Research-Intensive-University-Full-paper.pdf. Accessed 17 Oct 2018.

West, S. 2015. Negotiating Social Ecological Fit Through Knowledge Practice. Licentiate Thesis in Natural Resource Management, Stockholm University.

West, S., R. Beilin, and H. Wagenaar. 2019. Introducing a Practice Perspective on Monitoring for Adaptive Management. *People and Nature* 1: 387–405.

CHAPTER 5

Adaptive Doing in *Tarerer*/Kelly Swamp

In this chapter, we return to *Tarerer*/Kelly Swamp to illustrate the four phases of the adaptive doing process including our engagement in the agora. We use the Swamp that was described in Chapter 2 to demonstrate this process, drawing on the three different reframing tools—the 4 Is (interruption, interrogation, interaction, integration), assemblage, and the eternally unfolding present—that come from the social sciences and assist integration by offering different perspectives. These three reframing tools enable us to challenge assumptions and see SES in different ways to enhance our knowledge of the SES, integrate what we know and do together more effectively, and offer new insights that may lead to very different decision-making and action. For each of the three reframing tools, we outline what is involved in applying them to our interactions with the Swamp and the locals, what we examine with them, and what new shared understanding the reframing tool illuminated for us. To be clear, these three reframing tools are not the only way to achieve these outcomes. They serve as examples of what the social sciences can offer SES research and practice. We share the outcomes and reflections of our agora along the way as it aligns with the different phases of the adaptive doing process.

© The Author(s) 2020
A. Rawluk et al., *Practices in Social Ecological Research*,
https://doi.org/10.1007/978-3-030-31189-6_5

Phase A—The Trigger to Enter the Agora

In Phase A of the adaptive doing process some sort of novel experience triggers the need to enter the agora. For us, it was the observation that although researchers working in SES talked of social ecological systems as being integrated, the practices we observed were of frameworks and analysis that separated the social from the ecological.

Other triggers kept us returning to the agora—discovering more about *Tarerer*/Kelly Swamp, and reflecting on what reframing tools might be useful and why. Although we use *Tarerer*/Kelly Swamp as a case to demonstrate the process of adaptive doing in this final chapter, any SES could have been used. We have not chosen this case with the view to solve any of the pressing issues facing the Swamp. The Swamp served as a practical context in which to explore the ideas we present in this book. None of the authors had any significant knowledge of *Tarerer*/Kelly Swamp at the outset. We were able to build on local connections through our research networks and this enabled us to learn about *Tarerer*/Kelly Swamp. These different perspectives create many ways of seeing and experiencing the Swamp and each contains assumptions that privilege certain ways of doing and knowing and situate action for change. We suspect that all SES will similarly have multiple stories to share.

Preparations Within the Agora

In the interest of transparency, we suggest when forming a collaborative group in an agora to establish a common understanding of purpose (why are we here), an intention for the inquiry (what do we want to accomplish), in language that is accessible (being sure to monitor jargon), and agree to rules about trust and confidentiality, that are spelled out explicitly. In our experience, this has been an ongoing negotiation that has required continuous communication, redefinition, and clarification as we have refined our purpose and goals. We have found the adaptive doing process and working within the agora to be a time-intensive process, yet it comes with many rewards. We share both the challenges and insights below.

As part of the adaptive doing process, we also suggest that it is useful to clarify the differing appreciations of reality and multiple ways of knowing present in the agora, as well as being able to identify and analyse these as they emerge in data. The scholarship underpinning this is included in Chapter 3. We ensured all members could handle narrative data and conduct rudimentary thematic and narrative analysis by sharing skills and training each other. These thematic and narrative techniques are well-documented elsewhere (see Bryman 2015, for example). Narratives are useful as they offer a common way in which very different participants can engage with the issues under discussion. The narratives summarised in Chapter 2 were a synthesis of the impressions we each formed of the Swamp in our reading about and walking in the Swamp, as well as in talking with the local people. Our discussions and the training offered, enabled a collective understanding of the narratives that are evidence of our efforts to work in an interdisciplinary way. In our illustration of adaptive doing, we undertook an exploration of narratives as provided by local people in combination with the information we found through literature searches, as presented in Chapter 2.

Although power dynamics were present in our discussions in the agora, they did not reinforce any one way to do things, nor did they stop us from creating a transdisciplinary collaborative space that invited and valued the contributions of knowledge and practices of locals, some of whom did not identify with any particular discipline. William Blake (1967) made the point that 'without contraries there is no progression'. We progressed because of and despite the challenges we offered to each other. We grappled with ideas about agency; the ways in which choices and norms shaped or reinforced how we construct our material and intellectual framings of reality; the ways in which the agora can support dialogue around identifying or recasting our understanding and responsibilities. We illuminate the role of ethics and reflect on societal relationships that are revealed in the narrative contexts of the practices we analyse as we apply the three reframing tools. These kinds of critical reflections are part of an adaptive doing approach, which is a way of thinking, rather than a recipe for what to do. We describe here our experiences as participants in adaptive doing and in the agora.

One of our early discoveries was how on the surface, the Swamp did not seem a contentious site. Yet, we were to learn through listening, questioning, and uncovering the richness in the local stories and environmental histories, that there are many different understandings of the Swamp: a productive landscape, a critical bird habitat, a contested Indigenous site, and a recreational opportunity for cycling. Many of these understandings were not evident from our initial examination of the literature about the Swamp but became evident through our narrative analysis.

Phases A–C: Applying the 4 Is as a Reframing Tool to Structure and Function

The 4 Is reframing tool involves four aspects: interrogation, interruption, interaction, and integration (Beilin and Bender 2011) as introduced in Chapter 4. Drawing on the 4 Is, we identify the dominant components and relationships that shape the structure and function of an SES, and when this dominant understanding is disrupted, we draw on our social research practice of reading and listening more broadly to uncover multiple understandings of the components and relationships in the system. We then integrate the multiple understandings of the system around a shared concern for the resilience of the system and reflect on what new insights this provides. Our application of the 4 Is reframing tool to the *Tarerer*/Kelly Swamp SES occurs within our agora, beginning with **Interruption** in Phase A, through **Interrogation** and **Interaction** in Phase B and **Integration** in Phase C (see Fig. 5.1).

When we looked out over the Swamp, we all immediately saw the dominant system of agricultural production. We identified prominent anthropocentric components like animals, water, soil, economy, residential and recreational land use, as well as change in vegetation. In this dominant view of the system, we recognised the introduction of: grazing animals, such as sheep and cows; of hunting animals such as rabbits; and foxes as unwanted predators. We observed the migratory birds that visit the Swamp, and the vegetation components important for fodder and sand stabilisation. We became aware through the literature and narratives of the removal of indigenous vegetation, and that the land and soil is considered to be marginal. We saw how water had relationships that formed both benefits and challenges in terms of water quantity and quality, including the extraction of groundwater to enhance fodder production and efforts to drain and deviate the flow of water for the same goal. The economy was an important component of this system, with incomes from dairy farming on the decline. Recreation is a growing disruption to the agricultural dominance in the system, now that racehorses are being allowed to run on the dunes and walking and biking is permitted along the rail trail that cuts across the Swamp. Furthermore, when we started listening to the narrative of the artist, who spoke of a '*Tarerer*

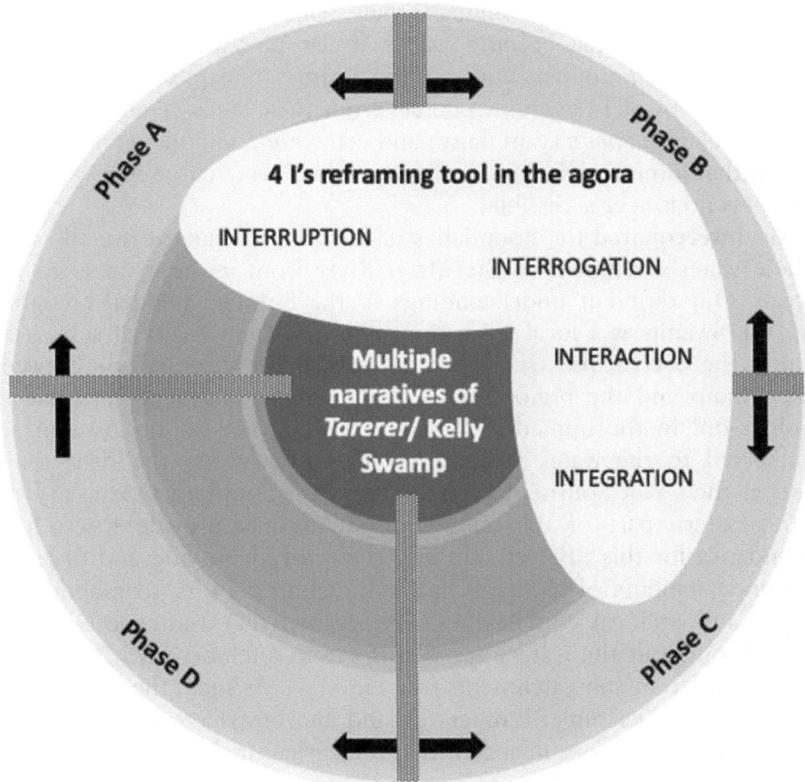

Fig. 5.1 Adapting doing in *Tarerer*/Kelly Swamp SES with the 4 Is reframing tool as an illustrative example used within the agora. Interruption, interrogation, interaction, and integration are shown to indicate when they might occur in relation to the adaptive doing phases. The other reframing tools of assemblage and the eternally unfolding present would be used during these same phases

festival', our understanding of the system was **interrupted**. *Tarerer* is the Indigenous, Gunditjmara name for this area where groups from across the region would gather. This was a crucial perspective on the system that was not expressed in describing the SES as simply an agricultural area or catchment basin. This was the first time that we heard the area being described as anything but 'Kelly Swamp' and realised there was a broader system framing that needed to be considered.

Seeing beyond the dominant view, shifts the SES components and relationships that become visible, it also placed a greater focus on Indigenous communities, culture, and history. It exposes the massacre and relocation of Indigenous peoples from their lands. Traditional food sources, like *murnong* (yam daisy) and eel, were components not identified in the dominant view of the system, which were culturally significant products no longer accessible.

We **interrogated** the boundaries of the SES, defining it overall as the whole water catchment of the Merri River from its headwaters to the ocean. Our different understandings of the Swamp included recognising the Swamp as a local subsystem within a larger regional subsystem within the overall SES (see Fig. 5.1), with many **interactions** between the Swamp and the region. While some of the local people spoke of 'goings-on' in the immediate Swamp, the activities in the Swamp are connected to the water flows across the estuary and the river to its exits at the Great Southern Ocean. The Great Southern Ocean and the south-eastern part of the continent of Australia provide macro-level boundaries for this SES, establishing a way of identifying and ordering our understanding of the landscape in the context of its geographic, climatic, and political complexities. We learned that Indigenous people lived throughout the waterways of this whole catchment, with different clans in different sub-catchments (see Fig. 2.5). As such, the structure of the system, for example, Indigenous land and broader water catchment, is far greater and more nuanced than the agricultural and recreation site depicted in the description in Chapter 2.

In considering the system more broadly, and from multiple viewpoints, many issues about its management emerged:

- agricultural production in the area is driven by markets that are international and national;
- conservation and recreation values tend to be isolated in the landscape so not clearly responsive to national or international agreements;
- there is no explicit Indigenous representation in the management of the area, although local Traditional Owners are consulted about some aspects of the public land;
- government is seen to be either removed and disinterested, or interfering and unrepresentative of citizen needs; and

- non-humans, such as migratory birds, are represented largely by citizens who volunteer their time, non-governmental organisations (NGOs), and some government agency staff, all of which are subject to funding cycles.

Delineating and describing these boundaries in the SES is an inherently social activity and the boundaries are socially constructed and flexible, with multiple framings and interpretations. These differences in view may only be initially exposed through interrogation.

We **integrate** multiple components and relationships (and boundaries) of the system that shape our understanding of its structure and function and orient around our shared concern for the resilience of the system. Resilience is often used as an indicator of the well-being of an SES, which is informed by its structure, function and ecological integrity (Walker and Salt 2006). We consider how a practice-oriented approach to observing SES, that draws on the multiple narratives in combination with the literature review, can provide different insights about the system.

Examining the system more broadly and focusing on (**integrating around**) the shared concern for resilience, multiple tensions were uncovered that were not apparent from the descriptions in Chapter 2. First, it showed that conflict between the first European settlers and Indigenous peoples has continued in the relationships between their descendants. The Traditional Owners now have limited or no access to lands where their ancestors would have once farmed and harvested foods. Farming in the Swamp is not resilient to environmental and economic pressures. Farmers have been attempting to modify their practices in response to their perception of land conditions by changing what they produce—potatoes, dairy, and most recently grazing and horse agistment. National and international economic and political tensions are making farming this land increasingly difficult, despite the benefits of the wetter areas, like *Tarerer*/Kelly Swamp offering grass much later into the dry season than in other areas. Policies could be targeted to either increase or decrease the viability of such farms. Recreational and residential land-use changes may also threaten conservation and heritage values. Changes in land use towards recreation, like the rail trail, may offer a compromise that alleviates agricultural pressure on the land, minimising disruption to wildlife by farm animals and motorised vehicles, while keeping human activity to negotiated boundaries. It may also create an opportunity for Indigenous peoples to gain more regular access and rights to these places. A broader understanding by the community of the water system, including the role of the dunes as 'sponges' that absorb and release water depending on how much water remains in *Tarerer*/Kelly Swamp, could shift how Parks Victoria and the farmers manage the timing of the opening of Rutledge cutting, which has many negative consequences for fish fry and the salinity of the land closer to the coast. Vegetation restoration is an activity that could be implemented more widely and more successfully if farmers, artists, and scientists worked together as they recognise that such vegetation can be beneficial to their animals.

In summary, using the 4 Is as a reframing tool within the adaptive doing phases A–C assisted comparison of our initial descriptions of the Swamp and the narratives as presented in Chapter 2. This comparison shows how drawing on the practice of listening to and including narratives revealed many additional issues that were not evident based on a literature review alone, one of which led to us reframing Kelly Swamp as *Tarerer*/Kelly Swamp. The literature review revealed an established binary, that of production versus conservation, and that there was a systems-view of the river as a catchment SES (the macro-system), among scientists. Whereas, the practice-oriented approach, using the 4 Is as a reframing tool exposed new tensions, such as between European settlers and Traditional Owners, and offered a number of different opportunities where further action could be taken to shift the state of the SES towards greater resilience, such as changes in access and shifts in vegetation management.

Phases A–C: Applying Assemblage as a Reframing Tool for Considering Relationships

A second way to challenge assumptions and more deeply understand aspects of an SES is by shifting the focus to relationships and drawing on the reframing tool of assemblage. We recognise the need to focus on particular elements of the SES in order to make sense of events or management decisions. These elements are likely to be the reason that participants would want to enter the agora, to discuss their experiences or interactions within the SES. Focusing on particular elements or relationships over others is a form of disruption (adaptive doing Phase A).

In this relational reframing tool, we draw on the social research practice of listening to how people construct their narratives and the relationships that they see and focus on in how they assemble their understanding of a place. This depiction of multiple and often competing assemblages has undertones of critical theory that can give rise to examinations of agency and power. In this illustration of a relational reframing tool, we focus on two example assemblages: migratory birds and agriculture. We orient these two assemblages around the shared concern for the presence and flow of water.

Narratives express different focal relationships or connections (Deleuze and Guattari 1987), or assemblages. For example, concerns by conservationists for the safety of migratory birds on the dunes, if dogs or racing horses are unrestricted, is an expression of different values and choices than those associated with the joy of riding a horse along the beach at sunset, or walking dogs in the early morning in these locations. Each of these perspectives is likely to be narrated with different endpoints and with different expectations of effect on the listeners. This points to the relational qualities of narratives. They are significantly about the connection between the narrator and the listener; as well as about the ways in which the narrators bring objects or elements 'into being'—make them visible to listeners and readers alike.

The initial description of the SES from Chapter 2, focusing on the biophysical and social components, expressed little of the relationships that existed between the biophysical and the social. Emphasising relationships extends the power or meaning of place. For example, to the local bird watchers the dunes are habitat for hooded plover shorebirds. The location is connected to a source of food (at the tide mark) and nesting (inter-tidal). For the horse rider, the beach is that wild, open, empty, long, unmarked expanse of sand. The beach as a location is entangled in images that connote space, for example, to run a horse. Connecting these words to these objects exposes emotion and intent. It also provides a richer picture of the site. This focus on connectedness and meaning resonates with our intention of acknowledging multiple ways of understanding how people live in and are entangled with the world around them. It also serves to show how articulating that entanglement can provide a deeper understanding of context and its translation into practice—into the things we do—our actions. Such ideas are expansive, inclusive, and embracing of human and non-humans. Relationships depend, at least in part, on the non-human which is central to the existence of land and water management practices and is always at the core of social ecological practices.

Migratory birds and their migration are an assemblage that provides a critically reflective counterpoint to the SES as previously described. Birds are represented in this assemblage by the international, national and local scientist voices who have studied their lifecycles, creating global documentation; and, by the regional presence or absence of these birds in the landscape—their physical manifestation as 'objects'. Knowledge about the birds builds relationships that depend on scientific trust, and academic connections for the scientists. These connections are tangible and intangible and can be interrupted by scientific error, funding deficits, or just not enough public support to be able to connect the science to changing lifecycle practices. Non-human interruption here may be in the form of new information that changes expectations of where birds will go or what they will eat or when they may come to the site. This information is noted and frequently generated in documents through scientific studies at a regional level, connecting science to community people involved in monitoring, as well as NGO groups.

The migratory flyways are technically acknowledged and protected by bilateral migratory bird accords. The relationship between the physical presence of the birds in the landscape and these documents demonstrates a strength of assemblage as a method of analysis. The documents are not visible in the region, and certainly the birds are not cognisant of their existence, but the documents have the potential to enable particular habitats to become a normative expectation for the birds and for the humans in these landscapes in which the birds are located. Swamps are habitats for birds, and are seen as areas that need to be changed to be productive from an agricultural point of view. We can see how the international and national accords interrupt the ability of public and private land managers to manage these areas for particular purposes.

Reflecting on what it means to protect these sites connects managers and birds to each other in tangible and intangible ways. Farmers may change their use of pesticides or decide to maintain particular habitats, and horse riders and dog walkers may change behaviour, to ensure the birds' safety. Public land manager oversight may involve baiting areas where foxes can be expected to interfere with breeding sites. The relationships above are based on previous disruptions to land and water management that conform to the expectations of international and national designation of sites as migratory bird flyways. Regional responses to those accords will create multiple connections between the people in these landscapes and the birds. The bilateral accord documents are, along with some of the scientists and some of the farmers and bird monitoring citizens, effectively the voices of the birds.

Thinking in terms of an assemblage centred on the migratory birds removes humans from the dominant position in the universe. Instead, it positions the non-human alongside, rather than beneath human priorities. Deliberately changing practices or managing landscapes for the benefit of the non-human may not be a common practice. Connecting pesticide use and seasonal planting regimes to habitat can begin with questions that examine the relationships between the different human representatives that manage land for production and those that manage for the birds.

A second assemblage observed in *Tarerer*/Kelly Swamp is around the relationships of the farmers. Farmers are in constant need of fresh water to produce fodder for their livestock, but not too much such that the plants drown. There is an uncertain supply of water across the Swamp due to changing rainfall patterns from the interior. The farmers relate the need for caution, given the expectations expressed by the national and international markets and the uncertainty of production regimes. There is a disconnect between what the market considers best practice and what farmers may be able to do. They receive predictions about market futures in at least two identified ways. One is through local agricultural production associations such as the dairy and cattle associations. These identify trends in the market as well as in the climate and act to promote best outcomes for the farmers based on latest education and other widely sourced information. Not all farmers belong to such organisations but many do, and others use social and work networks to hear the same information. These social network structures can create shadow networks, important to regional connectivity for many communities and citizens. The second influential connection between the markets and regional farmers are the trade and service companies. Farmers provide narratives that articulate their reliance on the advice of regional traders, identifying national and international buyers and best prices. Trade and service business relationships act as middlemen in the regional supply chain that tilts their narratives to external, global market imperatives, situated external to the region and the SES. This external connection also emphasises the fluidity of markets. This dynamism plays out as the temporal uncertainties of global–local market futures, it also connects the farmers' narratives to the raising of calves in less than ideal conditions—cold winter weather—for live sale overseas at an optimum weight. The farmers' narratives express their concern for the well-being of these calves, as well as the stress the farmers experience in achieving a suitable transport weight at the feedlot for these calves. Further connections are made to the condition of the pastures in wet winter weather and the longer timelines needed to overcome those resulting compacted paddocks in spring.

The relational analysis we undertook in applying assemblage has been particularly useful in giving weight to previously unseen and interdisciplinary aspects, as well as the obvious connections that may not be given attention. The assemblage tool revealed the multitude of interconnected relationships made tangible in a bird assemblage. From the perspective of those working at the micro level, such as focusing on a sense of obligation among individuals to leash their dogs on beaches (Williams et al. 2009), the assemblage tool expands the view and transcends disciplines, entwining the relationship of dogs and their owners with habitat relationships and hydrological ones, among many others, to build a much more holistic understanding. Additionally, the influence of cold weather and low feed stocks, due to a previous drought effects the ability of regional producers to meet the winter calf weight for overseas live transport in a current season. None of the connections—weather, feed, markets, animals, water supply—are reducible, or treated in isolation, though each may offer other ways of narrating the story. Farmers in the region may be encouraged to build winter feeding pads rather than keeping animals on the land; or to grow different crops in summer that would supplement normal feed regimes; and regional middlemen may seek transport services from nearer ports. Relational analysis through assemblage acknowledges dynamic fluctuations in the landscape, possible alliances between weather, crops, animals, and people, or between institutional structures such as market and water boards.

This consideration of the two assemblages could be seen as entirely in opposition, however, finding the overlap and the aspect of shared concern—namely the presence and flow of water—exposes an opportunity for working together. While the previous reframing tool of the 4 Is identified the tensions in the broader consideration of the SES, the focus here on assemblage identifies both the differences and the commonalities between focal relationships. Discussions of agriculture and migratory bird relationships can promote a common ground for concern and action in managing for the presence and flow of water.

PHASES A–C: APPLYING THE 'ETERNALLY UNFOLDING PRESENT' AS A REFRAMING TOOL

In the application of the third reframing tool, we turn to an exploration of the eternally unfolding present (EUP). We draw on a simplified adaptation based on West et al. (2019) to examine what we know and how practice evolves. Working in the EUP does not provide a recipe for what will emerge each time; but represents a state of awareness about the potency of local and deliberative action. It is empowering for those making decisions because it focuses on the artefacts and relationships that are involved in decision processes and the actions arising. As detailed in Chapter 4, the EUP enables the disruption of SES assumptions, and centres on timelessness and the irreducibility of the social ecological. It emphasises the importance of paying attention to the present moment and points to the importance of listening to and understanding each other's narratives. We draw on the social research practice of listening to people in the moment, to their intimate experience of being present. We provide two examples: an ecologist in a field watching birds; and a farmer enacting their practice of working the land.

An Ecologist's Narrative using the EUP: *The ecologist describes doing a bird count on a very still and sunny day. The birds they are looking for are on the endangered list and there is a sighting of them, so two department employees—the ecologist and a colleague—go to check on this reporting. The birds are there, at least they can be heard, recognisable by their calls to each other. In the first instance, two birds see the oncoming visitors and fly up, calling a warning. The others also rise up and retreat further down into the grasses. The ecologist experiences a strong sense of their fear. The humans stop where they are and wait for some time to pass, intent on calming the birds. The ecologist has experience of meditation practices, and says that he deliberately drew on that practice, initially to be patient, and more fully be present in the place. In narrating this to the authors, he changes tone, and describes a kind of trance-like affective experience. He senses the birds' presence and the birds' fear. He feels able to mentally reach out to the birds, seeing the landscape from a bird's eye view, and a sensory transformation. The ecologist goes deeply into self, seeking to mentally communicate with the spotter birds, seeking to assure them that he and colleague are not a threat,* *and will not harm them. They begin to walk quietly but openly towards the birds and are able to approach closer than the ecologist has ever been able to do before. The ecologist feels that the birds are aware and that the birds are allowing these humans closer—letting them approach. And then the birds call out and the small flock rise up. The moment is gone. The ecologist and his colleague have counted what they could, so their job is done; but the ecologist is profoundly affected.*

In an ideal situation, ecologists use bird hides to monitor birds with as little disturbance as possible. The recognition that the birds experience fear and fly off is not intended as an indictment of the ecologist's methods. Rather, it reflects the pressure that many who work in government departments and NGO with the goal of trying to protect habitat for endangered species experience. For that to happen, the species must be known to use the site, and therefore, be counted wherever they can be found to build the history for such protection claims. Further, in highly cleared landscapes such as in the study region, many indigenous species now use non-indigenous habitat and sightings are serendipitous rather than regular. This background explains the excitement of the quest, as well as the importance that it has for the ecologist in this story.

In **actionable understanding**, the authors begin by thinking through the ecologist's story, we hear the ecologist define his scientific work as requiring that he identify and count the birds at the site. This requires that he have monitoring sheets, camera, and binoculars as physical tools, but also a depth of knowledge about the bird habits, their preferred habitats (even though this is not one of those expected habitats) and their likely behaviours. The second part of this step is knowing how to access the site, given the conditions, to optimise success in identification and counting. To this end, the ecologist and colleague approach through the taller grasses and attempt to mask that approach with the vegetation. This approach tactic is discussed briefly before entering the site and their practice begins with walking as quietly as possible into the vegetation area.

When the birds are initially disturbed, they fly up and go deeper into the site. The ecologist becomes separated from his colleague in the sense that they are no longer side-by-side, and that they do not openly communicate. At this point, they draw on the way that they have been trained to monitor, their reason for being in the site. The everyday practice response might be to recall what is known about the birds' normal behaviour and characteristics from other sightings or textbook descriptions. The bird calls are different, for example, when there is disturbance, so initially they wait to hear a resumption of what might be called normal chatter. **Ongoing business** allows the ecologist to consider what routine actions or understandings of everyday practice will help in achieving the resumption of their identification task and subsequent count. It also provides a framing for recognition that the initial approach did not work, and that something else has to occur here to achieve their goal.

This leads to the ecologist changing his approach and being in the present: what we, in the agora, hear as the **eternally unfolding present**. He is able to draw on learnings from outside of his training as an ecologist. He recognises the importance of time in calming the situation for the birds. He has no hide to shelter in and must depend on just being in the landscape, being in that place in *basho*-like moments. The conditions have changed the context. What the authors heard in this narrative is that there is uncertainty at every level: the uncertain response of the birds—how close will they let the ecologist come; and the uncertainty associated with identifying and counting the birds—as the hoped-for endangered species. He is no longer there on his mission to identify and count and thereby consolidate habitat protection claims. Rather, as he describes his meditation, he is now supremely aware of the birds' fear and his goal is to be part of the place and minimise disturbance. He is a scientist and also someone, who through his meditative practice, becomes part of the place. In doing so he adapts his practice. He describes reaching out to the outlying birds in his mind. His awareness of their fear seems to trigger a communication between them. His conscientisation gives him a way of slowing the approach. He does not know how long this took because he is absorbed in being in that landscape.

As a reframing tool, the EUP challenges assumptions of language, social ecological order, and time in SES research. It can foreground the inherent uncertainty that is present in SES in terms of first the social ecological and second, of practice and knowledge. Listening to the relationships expressed in EUP narratives enables researchers to access the irreducibility of practice and knowledge. At other scales, the uncertainty in an SES can be categorised, ordered, and organised away, but at this focal level in the EUP, it is irreducible.

When focusing on the irreducible entanglement of the social ecological, in the example of the ecologist, there is no separation of the person mentally connecting with an inner, bird-sensing-person, and the birds in the quiet, sunny environment. In this example, the ecologist was expressing a dynamic relationship with what he was doing and it **changed his immediate practice**. In the transmission of this experience to others, it has the possibility of changing practice more widely. The bird monitoring in less than ideal situations is likely to be the norm for endangered species sightings. But an overall learning from the ecologist's experience can be about the embeddedness of species in the landscape, even highly altered landscapes such as in the study region. In a reality in which flexibility and adaptive thinking are certain to be required, the temptation is to reduce goals to achievable outcomes that are easily documented. But the ecologist's story is that these birds can only be found in the context of the changing landscapes that they too are adjusting to in order to survive. The landscape cannot be reduced to individual species counts. This is not such a palatable conclusion for funding in government programmes, nor is it likely to be a new finding for ecologists. But the embeddedness of species in landscapes affirms the importance of the irreducibility of social ecological connectedness.

A Farming Narrative using the EUP: *The farmers manage the flood plain alongside the estuary, Tarerer/Kelly Swamp and on to the west, where the estuary joins the sea, experience winter flooding. It is usual that their pastures are flooded from late July for a couple of months. In these times the farmers move their cattle to higher ground, some on neighbours' land. The loss of pasture, as it is understood in their narratives, is also exacerbated by the number of calves born at this time of year, and the pressure for sheltered spots, many of which are normally in the undulations leading to the flood plains. The bad weather, young calves, and wet paddocks create anxiety among the farmers. They want the water to abate quickly. In this narrative compilation, they take action to speed the draining of the flood plain in winter by cutting a drain across their properties. There is already an older drain in existence that was legitimated with the building of the Rail Trail, when the bridge and culvert over this area was publicly funded. The second and new drain rests a short distance away from the established one.*

In this narrative, farmers' experience of being good land managers helps them define their role in cutting the drain. Their aims consolidate the **actionable understanding** required to make the land accessible to cattle and usable for pasture. The farmer closer to the Merri River shows the authors how the flooding cuts almost in half his usable, warmest land areas in winter. He recalls terrible flood waters and the consequent losses he attributes to them. He demonstrates his management of the first drain, which he has recently cleaned, as evidence of its importance and contribution to the productivity of his property. The farmers on the western end of *Tarerer*/Kelly Swamp, explain how the flooding is only good for a short time, and then, the paddocks need to be returned to the rotation cycle. Their explanations for seasonal inundation, cycles of 'good' and 'too much' water, generate their surety of acting and their subsequent involvement in the practice of cutting the new drain.

Ongoing business returns us to the way in which the floodplain is managed over all the seasons. When the rest of the properties might be dried out in summer, these low-lying areas dry out last and have cattle and cows on them when there is little grazing elsewhere. The maintenance of the 'correct' amount of water in the floodplain is an undocumented knowing based on seasons of experience, and years of intergenerational stories that describe the quality and depth of water at different times. It is tacit and explicit knowing building up norms of expected management between the farmers across this area. Their roles as good land managers is partly bound up in knowing when to accept and when to combat those winter floods. Extreme weather can disrupt their intentions, prolonging drought until even the flood plain is browning in summer; deepening the water levels as interior storms are felt with the arrival of increased flooding in winter. These disruptions are understood as uncertain but needing to be guarded against with good decision options now.

The **eternally unfolding present** for the farmers on the day of the drain digging represents a culmination of all the seasons of variable weather and water collected into the now of acting. As land owners and managers they act to regulate the relationship between the land and their productivity. They act knowing that the drain will be there in dry times as well as in wet. The digging is done by three pieces of specialised machinery, shared by the farmers to construct their project. Their facility with the equipment brings actionable understanding and ongoing business into the EUP because it represents skills and job roles and all the taken for granted knowing that constitutes the location of the drain along the paddocks and, ultimately, its expected utility. They imagine the next flood and joke about the overcoming of some of its force, even as they recognise that in some years, it may make very little difference. It is evident from this excerpt that there are multiple timelines—the old drain which is evidence of some flood management success, the new drain, the intent of which is further flood regulation, and the cutting of the drain itself across the landscape. These are brought together in the present cutting decision. Then there are the cycles of weather that bring drought and flood but never exactly the same depth of water or lack of rain from year to year. Each drought, and each flood is experienced anew across the landscape and in the stories of how to manage it.

In the farmers example, the EUP as a reframing tool highlights entanglement of multiple timelines, integrating actors in a whole history. It enables the disruption of an SES as a snapshot in time, but as an overlay of multiple timelines and many, many small actions. This reframing tool does not try to force these different timelines into a single aligned timeframe. The EUP approach also permits connections across time through the 'whole history' that the narrators shaped and shared, and we hear how the farmers call on these memories and stories to act in the building of the second drain. The new drain is part of the history of the old drain just as they are all part of the management of this landscape. These relationships are not static. When we focus on the immediate moment of what an actor does and why, such as opening a drain, we understand that an actors' action is a part of an entangled knowledge-practice that is a product of the timelessness in which individuals are immersed. Starting with listening to and imagining the experience of how people do and what they know, without moving to a more abstract scale of categorising and ordering experience that offers the opportunity to judge it, offers a unique opportunity for finding commonalities among actors.

The EUP as a reframing tool enables researchers to challenge common assumptions in SES research by going to the core of SES uncertainties that are not often examined at broader focal levels (i.e., system and assemblage), such as irreducibility of the social ecological, dynamism of knowledge and practice, time, and power. Centring on such uncertainty and irreducibility in SES research can enable practitioners to find commonalities that might not have been otherwise identified. For example, both the ecologists and farmers are dynamically forming and shaping knowledge and practice in their experiences each moment, they are inherently existing in an irreducible social ecological system. When centring on their timeless experience in the EUP, actors are not seen as good or bad, right or wrong, but as individuals irreducibly connected through their actions to the social ecological. This irreducibility and timelessness could be central to identifying common ground and working together.

If the farmers had entered the agora with others who participated in this study, their management of the flood plain may have elicited responses from those concerned to prioritise habitat and conservation values, or those who are associated with river watchers or those who do not condone the cutting of the drains (because of the connectivity of the floodplain to other values in the area) even though the cutting is the property right of private landowners. In the agora, such discussions would emerge from narratives of flood plain use provided by others. It would allow other practices to emerge in the possible management of the water and the land. Acknowledging that uncertainty and irreducibility is ever-present can assist to make the uncertainty feel less of an obstacle and more of an opportunity for observing things differently and changing what we do.

Engaging Adaptive Doing Phase D: Reimaging and Returning to Practice

When we began writing this book, the authors thought of it as contributing to the toolboxes that would assist social ecological researchers and practitioners to bring together multiple knowledges, understandings, and practices for navigating collaboration, planning, decision-making, and management in social ecological systems.

The different critical reflections on the Swamp highlight the diversity of understandings of the landscape and SES. The irreducibility of elements within an SES suggests that the ideas and things that we choose to discuss are those that we bring into focus through a deliberative construction and/or emergence of relationships and/or spaces and all of these are part of the 'now'. They occur in the present. The other ideas and things we do not discuss in the agora, are still there, still exist, still are part of space and we can come back to them if it seems the paths we have set out on require such a revisit. No doors are closed. It is not necessary to achieve consensus, but a more comprehensive and shared understanding of an SES is developed through adaptive doing and critical reflection.

Regardless of the reframing tools used in adaptive doing, the participants who partake in critical reflection will have their knowing and practice changed. As authors in the agora, listening to and grappling with the narratives of informants, some of us understood new dimensions to the SES not observed by peering out of our temporary writing retreat accommodation window. The social and biophysical setting outlined in Chapter 2 foregrounds the conventional landscape binary between production and conservation. This contributes to the initial interpretations the authors understand as part of the local narratives we hear and analyse. Whereas we were able to learn from listening to and including the narratives of some of the local people, that there are additional ways of understanding practice that were not evident based on the literature review alone. The **4 Is** reframing tool, helped us to **interrupt** the Western naming of the swamp that we studied. We heard that it had a name that preceded Kelly Swamp: *Tarerer*. Interruption also assisted us to recognise the absence of the Traditional Owners in the management of this land and in our conversations for this book. We acknowledge that their inclusion would have enhanced our understanding of the Swamp, and that such involvement is an imperative to changing the historical construction of these landscapes, as well as our practices from here. Our **interrogation** through changing boundaries and scale revealed that there were multiple tensions that were not evident from the biophysical and social setting descriptions, such as the agency and effect of power. We also found that the setting of these boundaries is an inherently social activity with multiple framings and interpretations that may only be partially revealed through interrogation. From considering **interaction** and **integration** we better understand that there are challenges to agriculture as the dominant land use in this area. The short timelines of a more industrial agricultural practice on this land hides the previous incarnations of *murnong* (yam daisy), potatoes, dairy, and most recently grazing and horse agistment. Alongside, when we looked out at the view, observing pastoral land, water, dunes, cows and horses, and the occasional waterbird or bird of prey, we did not immediately observe the endangered orange-bellied parrot. Yet, when we considered the assemblage of the system as a point of critical reflection on relationality, we realised that the SES would appear entirely different if we were seeking orange-bellied-parrot relationships. Seeing the change in understanding based on the **assemblage** leads us to return to our practice clearer about the relationships we prioritise in decision-making, and those we knowingly or unknowingly ignore. Returning to our practice, we can think: whose stories are silenced in the dominant narrative or understanding of the SES? Considering the **eternally unfolding present**, we can see that the farmers are working to care for their calves and their land even as they are coerced by markets that limit their options and actions. Similarly, when we focus on the scale of an individual 'in the SES', existing now and connecting through the present moment to the past, present, and future, it leads us to critically reflect on what the unconscious biases, theories, or understandings are that we carry into our interpretation of landscape. We can seek out that which we cannot see. We understand that other possibilities for knowledge or practice are present when we detach from our theoretical framings.

Each of the authors are practitioners of research and teaching. Adaptive doing and critically reflecting on a social ecological system changed how we approach teaching with our adult students. We return to our teaching with a greater consideration of sharing the practice of critical reflection, providing deliberative attention to power in learning about contexts or phenomena and what is observed or silenced from discourses. The challenge and the discomfort of critical reflection and collaborative engagement remains. We are committed to foster learning experiences that enable open-minded and open-hearted engagement despite discomfort; and to asking students to always question and seek alternative understandings—in their studies and in their professional practices and everyday lives.

Conclusion: There Is No End to 'Adaptive Doing'

In the face of complex, interdisciplinary challenges—the 'wicked problems' in the world, this book argues for the primacy of practice to enable wholistic social ecological processes that will guide thinking and research and most importantly, recognise the power of 'adaptive doing'. We all spend our lives 'doing'. We observe, we feel, we experience, and our actions create common ground—entry points to exploring and recognising difference—because our differences lead to more robust pathways through our problems and issues.

In our illustration of adaptive doing, the three reframing tools applied to *Tarerer*/Kelly Swamp illustrate just three ways of reimagining our engagement with a social ecological system. We have used the *Tarerer*/Kelly Swamp example to illustrate the potential of the agora and the three tools for critical reflection. Combined, these offer other ways of seeing, imagining, and acting with collaborative and deliberative intent that will inform social ecological justice for humans and non-humans.

We opened the book with these three aims:

- to outline and demonstrate 'adaptive doing', a practice-oriented process for integrating research in SES, that is transparent, inclusive, and engaged;
- to demonstrate three reframing tools from the social sciences—the 4 Is, assemblage, and the eternally unfolding present—that assist SES researchers and practitioners to participate in 'adaptive doing'; and
- to overcome disciplinary silos by creating a platform that we call the agora, which creates a space where SES researchers and practitioners can participate in 'adaptive doing' to learn and improve SES practices and outcomes.

These five chapters provide:

- a book that supports taking action for social ecological justice;
- a synthesis of the current SES literature and background to what is needed to affect a practice focus that enables such action;
- an emphasis on the power and potential of interdisciplinary action to facilitate research and management; and,
- a kind of map, and a process-oriented guide.

Our rule has been to outline and demonstrate approaches that are transparent, inclusive, and facilitate engagement with contested subjects and landscapes. Further, to overcome disciplinary silos, we have argued for shifting positions, the reassembly of space and time, and the reimagining of boundaries and dialogues so that our discipline areas and ways of knowing (farming, ecology, volunteering, citizen...) can be heard and respected in the narratives that are told. We have deliberately attempted to provide space for those whose voices are missing from the enacting of 'adaptive doing' in *Tarerer*/Kelly Swamp. These are the Traditional Owners, other citizens who frequent these landscapes, and all the non-human species that inhabit or seasonally visit this system.

To later emphasise the repositioning offered by adaptive doing, we began Chapter 2 with a standard description of a site. We emphasised the geography and topography, the hydrology and the multiple views of the wetland landscape and its estuaries from the Merri River to the Great

Southern Ocean, separating the social and biophysical. This assembly of predominantly technical information was intended to help us locate the reader in the case site world. The establishing of landscape boundaries and multiple discipline views, as well as the utilitarian perspectives allowed us to remain as observers, looking into the locale.

This framing laid the foundation for Chapter 3, in which common assumptions in SES research were identified and the differences between social and ecological, nature and culture, location and place were examined, in the testament of the literature we reviewed and the gaps we identified. We argued for a way of integrating language to overcome limitations of what is implied and extend the potential intentions through words. The **social ecological** with no hyphen emerged as an example. Further, the literature assisted us to locate the social ecological within complex adaptive systems thinking. In this third chapter, we consider how uncertainty, non-linearity, emergence, and other complex adaptive system characteristics demand a change in the way SES practice is undertaken and importantly, how it is described to others. We reviewed ideas on determinism, the use of structure and function in social and biophysical writing and the importance of taking time to create a deliberative environment for dialogue. An emergent characteristic of this experience is that this literature focused our attention on how we learn and what we might do differently in our research on the ground and in our writing. We concluded this chapter with:

- critical reflection for action;
- critical reflection as a foundation for interdisciplinarity;
- critical reflections on ordering—and how we can disrupt conventional ordering (this leads to 4 Is as an approach);
- critical reflection on relationships via assemblage (this leads to assemblage as an approach);
- critical reflections on time (this leads to the eternally unfolding present as an approach); and,
- critical reflections on how learning brings these all together...single, double, and social learning.

Chapter 4 unleashes the excitement of adaptive doing as social ecological practice, reorienting conventional learning by emphasising multiple narratives of practice as a foundation for social learning. We describe and advocate for the power and potential of the agora—the space for changing practice—as a foundation for making space for agonistic views and recognition of previously hidden or ignored issues. The agora symbolises the centrality of multiple practices for robust change and the importance of drawing power into the open in an environment of trust. Consensus is not the necessary endpoint...rather, the intention that each narrative is heard and that all participants become genuinely engaged in changing their practices, emerges through the process.

In Chapter 4, Fig. 4.1 demonstrates the philosophical embeddedness of the ideas we analysed and synthesised in Chapter 3. We highlight how the use of narratives, the illumination of power, and multiple considerations of time, create space for shared concerns to be discussed. To support critical reflection and facilitate these discussions, we introduced three reframing tools: the 4 Is, assemblage, and the eternally unfolding present. They are useful in questioning the way we unthinkingly order the world and the assumptions and values that underpin conventional knowing.

This chapter provides practice examples of three reframing tools that can be used to create social ecological practice in our research, writing and fieldwork. The view in Chapter 2 was the status quo presentation of an SES. In this chapter, the critical reflection enabled by the reframing tools, provides a challenge to common assumptions in SES research, it triggers changes in view, and a respect for the agency of all participants including the non-human and those citizens who were not present at the time. And the purpose is to create an atmosphere in which we can all act for change through our 'adaptive doing'.

Our focus in this book has primarily been on research practice, which is where we tend to encounter social ecological challenges. However, concepts we have drawn upon of practice, experiential and social learning that underpin our understanding of adaptive doing are equally relevant to other participants in SES. For learning in any context, openness to variations in practice or experience, and tendencies to curiosity and reflection are important (Fazey et al. 2005). In organisations generally, social learning is enhanced by diversity and trust (Curseu and Schruijer 2010) and by flat power structures (Argyris 1976). Where we have drawn on methods for interdisciplinary engagement and academic analysis for our work within the agora, different methods could be appropriate for use with different groups of people. Depending on the participants and their context, the agora could emerge through a series of group discussions and interactions, beginning with verbal narratives of the SES, or with shared actions taking place in the landscape, depending on the participants and their context (Leeuwis 2004).

The approach we suggest in this book and the way of navigating through adaptive doing is not prescriptive or restrictive. We have written the book with particular backgrounds and interest in practice-based social research in addition to a range of disciplinary skills and knowledge. Those engaging with adaptive doing will be diverse and can potentially bring methods and aspects that we have not yet considered. What we emphasise in this book are principles and processes of critical reflection and consciencisation that would benefit social ecological research if viewed as a broad guide. They can be adapted with the introduction of different reframing tools and methods by those participating in the agora. We offer this book, adaptive doing and the agora as a catalyst to extend a new generation of social ecological research that sees participants honouring their disciplinary foundations, while being ready to change and collaborate within each new SES, and each new engagement: being able to act now, for social ecological recognition and change.

References

Argyris, C. 1976. Single-Loop and Double-Loop Models in Research on Decision-Making. *Administrative Science Quarterly* 21: 363–377.

Beilin, R., and H. Bender. 2011. Interruption, Interrogation, Integration and Interaction as Process: How PNS Informs Interdisciplinary Curriculum Design. *Futures* 43: 158–165.

Blake, W. 1967. *Songs of Innocence and of Experience: Shewing the Two Contrary States of the Human Soul, 1789–1794.* New York: Orion Press.

Bryman, A. 2015. *Social Research Methods.* Oxford, UK: Oxford University Press.

Curseu, P., and S. Schruijer. 2010. Does Conflict Shatter Trust or Does Trust Obliterate Conflict? Revising the Relationships Between Team Diversity, Conflict and Trust. *Group Dynamics: Theory, Research and Practice* 14: 66–79.

Deleuze, G., and F. Guattari. 1987. *A Thousand Plateaus.* Minneapolis, MN: University of Minnesota Press.

Fazey, I., J.A. Fazey, and D.M.A. Fazey. 2005. Learning More Effectively from Experience. *Ecology and Society* 10 (2): 4.

Leeuwis, C. 2004. *Communication for Rural Innovation.* Oxford, UK: Blackwell Science.

Walker, B., and D. Salt. 2006. *Resilience Thinking: Sustaining Ecosystems and People in a Changing World.* Washington, DC, USA: Island Press.

West, S., R. Beilin, H. Wagenaar, and C. Watkins. 2019. Introducing a Practice Perspective on Monitoring for Adaptive Management. *People and Nature* 1 (3): 387–405.

Williams, K.J.H., M.A. Weston, S. Henry, and G.S. Maguire. 2009. Birds and Beaches, Dogs and Leashes: Dog Owners' Sense of Obligation to Leash Dogs on Beaches in Victoria, Australia. *Human Dimensions of Wildlife* 14 (2): 89–101.

Literature Cited

Alaimo, S. 2008. Trans-corporeal Feminisms and the Ethical Space of Nature. In *Material Feminisms*, ed. S. Alaimo and S. Hekman, 237–264. Bloomington, IN: Indiana University Press.

Allison, H., and R. Hobbs. 2004. Resilience, Adaptive Capacity, and the "Lock-in Trap" of the Western Australian Agricultural Region. *Ecology and Society* 9 (1): 3.

Anderies, J.M., P. Ryan, and B. Walker. 2006. Loss of Resilience, Crisis, and Institutional Change: Lessons from an Intensive Agricultural System in Southeastern Australia. *Ecosystems* 9 (6): 865–878.

Anderson, N.M., R.M. Ford, and K.J.H. Williams. 2017. Contested Beliefs About Land-Use are Associated with Divergent Representations of a Rural Landscape as Place. *Landscape and Urban Planning* 167: 75–89.

Anderson, B., M. Keanes, C. McFarlane, and D. Swanton. 2012. On Assemblages and Geography. *Dialogues in Human Geography* 2 (2): 171–189.

Anderson, B., and C. McFarlane. 2011. Assemblage and Geography. *Area* 43 (2): 124–127.

Angelstam, P., M. Elbakidze, R. Axelsson, N.E. Koch, T.I. Tyupenko, A.N. Mariev, and L. Myhrman. 2013. Knowledge Production and Learning for Sustainable Landscapes: Forewords by the Researchers and Stakeholders. *AMBIO* 42 (2): 111–115.

Arendt, H. 1961. The Concept of History. In *Between Past and Future*. New York, NY: Viking Press.

Arendt, H. 1998/1958. *The Human Condition*, 2nd ed. Chicago and London: Chicago University Press.

Argyris, C. 1976. Single-Loop and Double-Loop Models in Research on Decision-Making. *Administrative Science Quarterly* 21: 363–377.

Armitage, D., M. Marschke, and R. Plummer. 2008. Adaptive Co-management and the Paradox of Learning. *Global Environmental Change* 18: 86–98.

Armitage, D., R. Plummer, F. Berkes, R.I. Arthur, A.T. Charles, I.J. Davidson-Hunt, A.P. Diduck, N.C. Doubleday, D.S. Johnson, M. Marshcke, P. McConney, E.W. Pinkerton, and E.K. Wollenberg. 2009. Adaptive Co-management for Social-Ecological Complexity. *Frontiers in Ecology and the Environment* 7 (2): 95–102.

Atifi, H., and M. Marcoccia. 2017. Exploring the Role of Viewers' Tweets in French TV Political Programs: Social TV as a New Agora? *Discourse, Context & Media* 19: 31–38.

Atkinson, M. 2011. Lindblom's Lament: Incrementalism and the Persistent Pull of the *status quo*. *Policy and Society* 30 (1): 9–18.

Aton, A. 2017. Earth Almost Certain to Warm by 2 Degrees Celsius. *Climate Wire*. https://www.scientificamerican.com/article/earth-almost-certain-to-warm-by-2-degrees-celsius/.

Ayre, M., and R. Nettle. 2015. Doing Integration in Catchment Management Research: Insights into a Dynamic Learning Process. *Environmental Science and Policy* 47: 18–31.

Bandura, A. 1971. *Social Learning Theory*. New York: General Learning Press.

Barad, K. 2007. *Meeting the Universe Halfway: Quantum Physics and the Entanglement of Matter and Meaning*. Durham, NC: Duke University Press.

Begon, M., J.L. Harper, and C.R. Townsend. 1990. *Ecology: Individuals, Populations and Communities*, 2nd ed. Cambridge, MA: Blackwell Scientific.

Beilin, R., and H. Bender. 2011. Interruption, Interrogation, Integration and Interaction as Process: How PNS Informs Interdisciplinary Curriculum Design. *Futures* 43: 158–165.

Berkes, F., J. Colding, and C. Folke (eds.). 2003. *Navigating Social-Ecological Systems: Building Resilience for Complexity and Change*. Cambridge, UK: Cambridge University Press.

Berkes, F., and C. Folke (eds.). 1998. *Linking Social and Ecological Systems: Management Practices and Social Mechanisms for Building Resilience*. Cambridge, UK: Cambridge University Press.

Berkes, F., and D. Jolly. 2001. Adapting to Climate Change: Social-Ecological Resilience in a Canadian Western Arctic Community. *Conservation Ecology* 5 (2): 18.

Berkes, F., and H. Ross. 2013. Community Resilience: Toward an Integrated Approach. *Society and Natural Resources* 26 (1): 5–20.

Bhattarai, B., R. Beilin, and R. Ford. 2015. Gender, Agrobiodiversity and Climate Change: A Study of Adaptation Practices in the Nepal Himalayas. *World Development* 70: 122–132.

Binder, C.R., J. Hinkel, P.W.G. Bots, and C. Pahl-Wostl. 2013. Comparison of Frameworks for Analyzing Social-Ecological Systems. *Ecology and Society* 18 (4): 26.

Blake, W. 1967. *Songs of Innocence and of Experience: Shewing the Two Contrary States of the Human Soul, 1789–1794.* New York: Orion Press.

Blaser, M. 2013. Ontological Conflicts and the Stories of Peoples in Spite of Europe. *Current Anthropology* 54 (5): 547–568.

Borrini-Feyerabend, G., M. Taghifavar, J.C. Nguinguiri, and V. Ndangang. 2000. *Environmental Management Co-management of Natural Resources Organising, Negotiating and Learning-by-Doing,* Eschborn, IUCN, GTZ.

Bourdieu, P. 1977. *Outline of Theory of Practice.* Cambridge, UK: Cambridge University Press.

Bracken, L.J., and E.A. Oughton. 2006. What Do You Mean? The Importance of Language in Developing Interdisciplinary Research. *Transactions of the Institute of British Geographers* 31 (3): 371–382.

Briassoulis, H. 2015. The Socio-Ecological Fit of Human Responses to Environmental Degradation: An Integrated Assessment Methodology. *Environmental Management* 56 (6): 1448–1466.

Briassoulis, H. 2017. Response Assemblages and Their Socioecological Fit: Conceptualizing Human Responses to Environmental Degradation. *Dialogues in Human Geography* 7 (2): 166–185.

Builth, H. 2009. Intangible Heritage of Indigenous Australians: A Victorian Example. *Historic Environment* 22 (3): 24–31.

Burton, D. 1982. Through Glass Darkly, Through Dark Glasses. In *Language and Literature,* ed. R. Carter. London: Allen and Unwin.

Camp, J.Mc.K., and C.A. Mauzy. 2009. *The Athenian Agora: New Perspectives on an Ancient Site.* Mainz am Rhein: Verlag Philipp von Zabern.

Capra, F., and P.L. Luisi. 2014. *The Systems View of Life: A Unifying Vision.* Cambridge: Cambridge University Press.

Chamala, C., and P. Mortiss. 1990. *Working Together for Landcare: Group Management Skills and Strategies.* Brisbane: Australian Academic Press.

Chaudhary, S., A. McGregor, D. Houston, and N. Chettri. 2015. The Evolution of Ecosystem Services: A Time Series and Discourse-Centered Analysis. *Environmental Science and Policy* 54: 25–34.

Chia, R. 2003. From Knowledge-Creation to the Perfection of Action: Tao, Basho and Pure Experience as the Ultimate Ground of Knowing. *Human Relations* 56 (8): 953–981.

Christensen, L., and N. Krogman. 2012. Social Thresholds and Their Translation into Social-Ecological Management Practices. *Ecology and Society* 17 (1): 5.

Clark, I.D. 1990. *Aboriginal Languages and Clans: An Historical Atlas of Western and Central Victoria 1800–1900.* Monash University: Melbourne Australia.

Clark, I.D. 1995. *Scars on the Landscape: A Register of Massacre Sites in Western Victoria 1803–1859.* Canberra: Aboriginal Studies Press.

Climate Reality Project. 2017. Why People Ignore the Science Behind the Climate Crisis (And What You Can Do). https://www.climaterealityproject.org/blog/why-people-ignore-science-behind-climate-crisis-and-what-you-can-do.

Crosby, A. 1986. *Ecological Imperialism: The Biological Expansion of Europe, 900–1900.* New York: Cambridge University Press.

Collard, R.-C., L.M. Harris, N. Heynen, and L. Mehta. 2018. The Antinomies of Nature and Space. *Environment and Planning E: Nature and Space* 1 (1–2): 3–24.

Constant, E., and A. Pickering. 1997. The Mangle of Practice: Time, Agency, and Science. *Technology and Culture* 38 (3): 815.

Cook, S.D.M., and H. Wagenaar. 2012. Navigating the Eternally Unfolding Present: Toward an Epistemology of Practice. *The American Review of Public Administration* 42 (1): 3–38.

Cooke, B., S. West, and W.J. Boonstra. 2016. Dwelling in the Biosphere: Exploring an Embodied Human-Environment Connection in Resilience Thinking. *Sustainability Science* 11: 831–843.

Cumming, G. 2014. Theoretical Frameworks for the Analysis of Social-Ecological Systems. In *Social-Ecological Systems in Transition*, ed. S. Sakai and C. Umetsu. Otsu, Japan: Springer.

Curseu, P., and S. Schruijer. 2010. Does Conflict Shatter Trust or Does Trust Obliterate Conflict? Revising the Relationships Between Team Diversity, Conflict and Trust. *Group Dynamics: Theory, Research and Practice* 14: 66–79.

Czarniawska, B. 2004. *Narratives in Social Science Research.* London: Sage.

Daly, M. 1978. *Gyn/Ecology: The Metaethics of Radical Feminism.* Boston: Beacon Press.

Datta, R. 2015. A Relational Theoretical Framework and Meanings of Land, Nature, and Sustainability for Research with Indigenous Communities. *Local Environment* 20 (1): 102–113.

DeLanda, M. 2006. *A New Philosophy of Society: Assemblage Theory and Social Complexity.* London: Continuum.

Deleuze, G., and F. Guattari. 1987. *A Thousand Plateaus.* Minneapolis, MN: University of Minnesota Press.

Desouza, K., and A. Bhagwatwar. 2014. Technology-Enabled Participatory Platforms for Civic Engagement: The Case of U.S. Cities. *Journal of Urban Technology* 21 (4): 25.

Dewey, J. 1929. *Experience and Nature.* New York, NY: W. W. Norton.

Dewey, J. 1954. *The Public and Its Problems.* Denver, CO: Alan Swallow (Originally published in 1927).

Dewey, J. 1988. *Reconstruction in Philosophy: Middle Works 1899–1924*, vol. 12. Carbondale, IL: Southern Illinois University Press (Original work published 1920).

Dixon, T.H. 2017. Curbing Climate Change: Why It's So Hard to Act in Time. *The Conversation*. http://theconversation.com/curbing-climate-change-why-its-so-hard-to-act-in-time-80117.

Dorward, A.R. 2014. Livelisystems: A Conceptual Framework Integrating Social, Ecosystem, Development, and Evolutionary Theory. *Ecology and Society* 19 (2): 44.

Doyle, H. 2006. *Moyne Heritage Study Volume 2: Environmental History*. Prepared for Moyne Shire Council in association with Context Pty Ltd.

Dunn, G., R.R. Brown, J.J. Bos, and K. Bakker. 2017. Standing on the Shoulders of Giants: Understanding Changes in Urban Water Practice Through the Lens of Complexity Science. *Urban Water Journal* 14 (7): 758–767.

Durkheim, E. 1982. *The Rules of Sociological Method*, ed. S. Lukes. London, UK: The Free Press.

Eckley, L., N. Hovelsrud-Broda, G. Kasperson, J.X. Kasperson, R.E. Luers, A. Martello, S. Mathiesen, M.L. Naylor, R. Polsky, C. Pulsipher, A. Schiller, A. Selin, and H. Tyler. 2003. Illustrating the Coupled Human-Environment System for Vulnerability Analysis: Three Case Studies. *Proceedings of the National Academy of Science USA* 100: 8080–8085.

Engel, P.G.H., and M. Saloman. 1997. *Facilitating Innovation for Development: A RAAKS Resource Box*. Amsterdam: KIT Pub.

Environment Protection Authority (EPA). 2004. *Environmental Audit: Merri River Estuary Findings and Recommendations*. Melbourne: Environment Protection Authority.

Environment Protection Authority (EPA). 2011. *How Will Climate Change Affect Victorian Estuaries?* Melbourne, VIC: Environment Protection Authority.

Fabinyi, M., L. Evans, and S.J. Foale. 2014. Social-Ecological Systems, Social Diversity, and Power: Insights from Anthropology and Political Ecology. *Ecology and Society* 19 (4): 28–40.

Farms, Rivers and Markets Project. 2012. *Farms, Rivers and Markets: Overview Report*. VIC. https://industry.eng.unimelb.edu.au/__data/assets/pdf_file/0007/2845609/farms-rivers-and-markets.pdf. Accessed 8 Oct 2018.

Fazey, I., J.A. Fazey, and D.M.A. Fazey. 2005. Learning More Effectively from Experience. *Ecology and Society* 10 (2): 4.

Fenton, C. 2005. *Hydrodynamics and Nutrient Status of the Fitzroy River Estuary—Progress Report 5*. Warrnambool: Deakin University.

Ferreira, F.S. 2017. Critical Sustainability Studies: A Holistic and Visionary Conception of Socio-Ecological Conscientization. *Journal of Sustainability Education* 13: 1–22.

Fine, G.A., and T. McDonnell. 2007. Erasing the Brown Scare: Referential Afterlife and the Power of Memory Templates. *Social Problems* 54 (2): 170–187.

Flyvbjerg, B. 1998. *Rationality & Power: Democracy in Practice.* Chicago, USA: University of Chicago Press.

Folke, C. 2006. Resilience: The Emergence of a Perspective for Social-Ecological Systems Analyses. *Global Environmental Change* 16: 253–267.

Foucault, M. 1994/1972. *Archaeology of Knowledge.* New York, USA: Tavistock.

France, B., S. Birdsall, and L. Simonneaux. 2017. Analysing the Multiplicity of Voices in the Agora: Using Actor-Network Theory to Unravel a Complex Issue. *International Journal of Science Education, Part B* 7 (4): 323–340.

Fratini, C.F., M. Elle, M.B. Jensen, and P.S. Mikkelsen. 2012. A Conceptual Framework for Addressing Complexity and Unfolding Transition Dynamics When Developing Sustainable Adaptation Strategies in Urban Water Management. *Water Science & Technology* 66 (11): 2393–2401.

Freire, P. 1970. *Pedagogy of the Oppressed.* New York, USA: The Continuum International Publishing Group Inc.

Friis, C., and J.O. Nielsen. 2017. On the System: Boundary Choices, Implications, and Solutions in Telecoupling Land Use Change Research. *Sustainability* 9 (974): 1–20.

Funtowicz, S., and J.R. Ravetz. 1994. The Worth of a Songbird: Ecological Economics as a Post-normal Science. *Ecological Economics* 10: 197–207.

Galafassi, D., T. Daw, L. Munyi, K. Brown, C. Barnaud, and I. Fazey. 2017. Learning About Social-Ecological Trade-Offs. *Ecology and Society* 22 (1): 2.

Gallopín, G.C. 1991. Human Dimensions of Global Change: Linking the Global and the Local Processes. *Global Environmental Change* XLIII (4): 3–17.

Gallopín, G.C., S. Funtowicz, M. O'Connor, and J. Ravetz. 2001. Science for the Twenty-First Century: From Social Contract to the Scientific Core. *International Social Science Journal* 53 (168): 219–229.

Gill, E.D. 1954. Aboriginal Kitchen Middens and Marine Shell Beds. *Mankind* 4 (6): 249–254.

Gill, E.D. 1978. *Quantification of Coastal Processes as a Basis for Coastal Management and Engineering.* CSIRO.

Gill, E.D. 1984. *Coastal Processes and the Sanding of Warrnambool Harbour. Definition of Coastal Processes, Quantification of Sand Erosion/Deposition, and the Reason for the Sanding Up of the Harbour at Warrnambool, S.W. Victoria, Australia* (editor). Warrnambool Institute Press.

Glenelg Hopkins Catchment Management Authority (GHCMA). 2008. *Merri Estuary Management Plan.* Hamilton, VIC.

Gunderson, L., and C. Holling. 2002. *Panarchy: Understanding Transformations in Human and Natural Systems.* Washington, DC, USA: Island Press.

Grove, K. 2014. Agency, Affect and the Immunological Politics of Disaster Resilience. *Environment and Planning D: Society and Space.* 32 (2): 240–256.

Grove, J. 2016. Response to Jedediah Purdy. *Forum: The New Nature, Boston Review*, January 11.

Haidt, J., and C. Joseph. 2004. Intuitive Ethics: How Innately Prepared Intuitions Generate Culturally Variable Virtues. *Daedalus* 133 (Fall): 55–67.

Haraway, D. 1988. Situated Knowledges: The Science Question in Feminism and the Privilege of Partial Perspective. *Feminist Studies* 14 (3): 575–599.

Haraway, D.J. 2008. *When Species Meet*. Minneapolis and London: University of Minnesota Press.

Haraway, D., and M. Kenney. 2015. Anthropocene, Capitalocene, Chthulhocene. In *Art in the Anthropocene: Encounters Among Aesthetics, Politics, Environments and Epistemologies*, ed. Heather Davis and Etienne Turpin, 255–270. London: Open Humanities Press.

Hatt, K. 2013. Social Attractors: A Proposal to Enhance "Resilience Thinking" About the Social. *Society & Natural Resources* 26 (1): 30–43.

Head, L. 2016. *Hope and Grief in the Anthropocene*. London: Routledge.

Heathcote, J., and S. Maroske. 1996. Drifting Sand and Marram Grass. *The Victorian Naturalist* 113 (1): 13.

Heemskerk, M., K. Wilson, and M. Pavao- Zuckerman. 2003. Conceptual Models as Tools for Communication Across Disciplines. *Conservation Ecology* 7: 8–12.

Herrero-Jáuregui, C., C. Arnaiz-Schmitz, M. Fernanda Reyes, M. Telesnicki, I. Agramonte, M.H. Easdale, M. Fe Schmitz, M. Aguiar, A. Gómez-Sal, and C. Montes. 2018. What Do We Talk about When We Talk about Social-Ecological Systems? *A Literature Review Sustainability* 10 (2950): 1–14.

Holling, C.S., L.H. Gunderson, and G.D. Peterson. 2002. Sustainability and panarchies. In *Panarchy: Understanding Transformations in Human and Natural Systems*, ed. L.H. Gunderson and C.S. Holling, 63–102. Washington, DC, USA: Island Press.

Ilicheva, A. 2010. Wild in the City: Past, Present, and Future. *Yearbook of the Association of Pacific Coast Geographers* 72: 56–72.

Ingalls, M.L. 2017. Not Just Another Variable: Untangling the Spatialities of Power in Social-Ecological Systems. *Ecology and Society* 22 (3): 20.

Ison, R., C. Blackmore, and B.L. Iaquinto. 2013. Towards Systemic and Adaptive Governance: Exploring the Revealing and Concealing Aspects of Contemporary Social-Learning Metaphors. *Ecological Economics* 87: 34–42.

Jones, O. 2009. After Nature: Entangled Worlds. In *A Companion to Environmental Geography*, ed. N. Castree et al. Oxford: Wiley-Blackwell.

Jones, M. 2014. Chantal Mouffe's Agonistic Project: Passions and Participation. *Parallax* 20 (2): 14–30.

Kearney, A.R. 1994. Understanding Global Change: A Cognitive Perspective on Communicating Through Stories. *Climate Change* 27: 419–441.

Keightley, E., and M. Pickering. 2012. *The Mnemonic Imagination: Remembering as Creative Practice.* Basingstoke, UK: Palgrave Macmillan.

Kharazzi, A., B.D. Fath, and H. Katzmair. 2016. Advancing Empirical Approaches to the Concept of Resilience: A Critical Examination of Panarchy. *Ecological Information, and Statistical Evidence Sustainability* 8 (9): 935.

Kolb, D. 1984. *Experiential Learning: Experience as the Source of Learning and Development.* Englewood Cliffs, NJ: Prentice Hall.

Krishnan, A. 2009. What Are Academic Disciplines? Some Observations on the Disciplinarity vs. Interdisciplinarity Debate. ESRC National Centre for Research Methods. University of Southhampton. Retrieved from http://www.forschungsnetzwerk.at/downloadpub/what_are_academic_disciplines2009.pdf.

Latour, B. 1986. The Powers of Association. In *Power, Action and Belief,* ed. John Law, 264–280. London: Routledge and Kegan Paul.

Law, J. 1999. After ANT: Complexity, Naming and Topology. *The Sociological Review* 47 (1): 1–14.

Lazarus, E.D. 2017. Toward a Global Classification of Coastal Anthromes. *Land* 6 (13): 1–27.

Lebel, L., E. Nikitina, C. Pahl-Wostl, and C. Kneiper. 2013. Institutional Fit and River Basin Governance: A New Approach Using Multiple Composite Measures. *Ecology and Society* 18 (1): 1.

Leduc Browne, P. 2018. Reification and Passivity in the Face of Climate Change. *European Journal of Social Theory* 21 (4): 435–452.

Leeuwis, C. 2004. *Communication for Rural Innovation.* Oxford, UK: Blackwell Science.

Lejano, R.P., and R. Shankar. 2013. The Contextualist Turn and Schematics of Institutional Fit: Theory and a Case Study from Southern India. *Policy Sciences* 46: 83–102.

Liu, J. 2017. Integration Across a Metacoupled World. *Ecology and Society* 22 (4): 29.

Liu, J., T. Dietz, S.R. Carpenter, M. Alberti, C. Folke, E. Moran, A.N. Pell, P. Deadman, T. Kratz, J. Lubchenco, E. Ostrom, Z. Ouyang, W. Provencher, C.L. Redman, S.H. Schneider, and W.W. Taylor. 2007. *Complexity of Coupled Human and Natural Systems Science* 317 (5844): 1513–1516.

Luthe, T. 2017. Success in Transdisciplinary Sustainability Research. *Sustainability (Switzerland)* 9 (1): 71.

Lyon, C., and J.R. Parkins. 2013. Toward a Social Theory of Resilience: Social Systems, Cultural Systems, and Collective Action in Transitioning Forest-Based Communities. *Rural Sociology* 78 (4): 528–549.

MacMynowski, D.P. 2007. Pausing at the Brink of Interdisciplinarity: Power and Knowledge at the Meeting of Social and Biophysical Science. *Ecology and Society* 12 (1): 20.

Malm, A. 2013. The Origins of Fossil Capital: From Water to Steam in the British Cotton Industry. *Historical Materialism* 21 (1): 15–68.

Marsh, D., S.A. Ercan, and P. Furlong. 2017. A Skin Not a Sweater: Ontology and Epistemology in Political Science. In *Theory and Methods in Political Science*, ed. V. Lowndes, D. Marsh, and G. Stoker. London: Macmillan International Higher Education.

McDonald, D., G. Bammer, and P. Deane. 2009. *Research Integration Using Dialogue Methods.* Canberra: ANU Press.

McGregor, J. 1995. *The Merri River—An Environmental Audit.* Warrnambool: Deakin University.

McKeon, M. 1994. The Origins of Interdisciplinary Studies. *Eighteenth-Century Studies* 28 (1): 17–28.

McNiven, I.J., J. Crouch, T. Richards, K. Sniderman, N. Dolby, and Gunditj Mirring Traditional Owners Aboriginal Corporation. 2015. Phased Redevelopment of an Ancient Gunditjmara Fish Trap over the Past 800 Years: Muldoons Trap Complex, Lake Condah, Southwestern Victoria. *Australian Archaeology* 81: 44–58.

Meadows, D.H. 1982. Sustaining Tropical Forest Resources: A Systems Approach. Unpublished Manuscript Obtained from Jennifer Robinson. Resource Policy Center, Dartmouth College, Hanover, NH, USA.

Mezirow, J. 1997. Transformative Learning: Theory to Practice. *New Directions for Adult and Continuing Education* 74: 5–12.

Mezirow, J. 1998. On Critical Reflection. *Adult Education Quarterly* 48 (3): 185.

Mirzoeff, N. 2016. It's Not the Anthropocene, It's the White Supremacy Scene, or, the Geological Color Line. In *After Extinction*, ed. Richard Grusin. Minneapolis: University of Minnesota Press.

Mol, A. 1999. *The Body Multiple: Artherosclerosis in Practice.* Durham, NC: Duke University Press.

Mondon, J., J. Sherwood, and F. Chandler. 2003. *Western Victorian Estuaries Classification Project.* Warrnambool: Deakin University.

Moore, J. 2015. *Capitalism in the Web of Life: Ecology and the Accumulation of Capital.* London: Verso.

Mouffe, C. 2000. *The Democratic Paradox.* London and New York: Verso.

Newell, B., C. Crumley, N. Hassan, E. Lambin, C. Pahl-Wostl, A. Underdal, and R. Wasson. 2005. A Conceptual Template for Integrative Human-Environment Research. *Global Environmental Change* 15: 299–307.

Nonaka, I., and N. Konno. 1998. The Concept of 'Ba': Building a Foundation for Knowledge Creation. *California Management Review* 40 (Spring): 40–54.

Odum, E. 1975. *Ecology: The Link Between the Natural and the Social Sciences*, 2nd ed. New York: Holt, Rinehart and Winston.

O'Malley, M. 2008. "Everything is Everywhere: But the Environment Selects": Ubiquitous Distribution and Ecological Determinism in Microbial Biogeography. *Studies in History and Philosophy of Biological and Biomedical Sciences* 39 (2008): 314–325.

Ostrom, E. 1990. *Governing the Commons: The Evolution of Institutions for Collective Action.* Cambridge, UK: Cambridge University Press.

Ostrom, E. 2007. A Diagnostic Approach for Going Beyond Panaceas. *PNAS* 104 (39): 15181–15187.

Ostrom, E. 2009. A General Framework for Analyzing Sustainability of Social-Ecological Systems. *Science* 325 (5939): 419–422.

Pahl-Wostl, C. 2009. A Conceptual Framework for Analysing Adaptive Capacity and Multi-level Learning Processes in Resource Governance Regimes. *Global Environmental Change* 19: 354–365.

Pahl-Wostl, C., L. Lebel, C. Knieper, and E. Nikitina. 2012. From Applying Panaceas to Mastering Complexity: Toward Adaptive Water Governance in River Basins. *Environmental Science and Policy* 23: 24–34.

Pascoe, B. 2014. *Dark Emu: Aboriginal Australia and the Birth of Agriculture.* Broome, WA: Magabala Books Aboriginal Corporation.

Pelling, M., and D. Manuel-Navarrete. 2011. From Resilience to Transformation: The Adaptive Cycle in Two Mexican Urban Centers. *Ecology and Society* 16 (2): 11–22.

Perz, S.G. 2019. Crossing Boundaries for Collaboration in Comparative Perspective: Key Insights, Lessons Learned, and Recommendations for Future Practice. In *Collaboration Across Boundaries for Social-Ecological Systems Science*, ed. S. Perz. London, UK: Palgrave Macmillan.

Pickering, A. 1992. From Science as Knowledge to Science as Practice. In *Science as Practice and Culture*, 1–28. Chicago, IL: University of Chicago Press.

Pickering, A. 2010. *The Mangle of Practice: Time, Agency, and Science.* Chicago: University of Chicago Press.

Phoenix, C., N.J. Osborne, C. Redshaw, R. Moran, W. Stahl-Timmins, M.H. Depledge, L.E. Fleming, and B.W. Wheeler. 2013. Paradigmatic Approaches to Studying Environment and Human Health: (Forgotten) Implications for Interdisciplinary Research. *Environmental Science and Policy* 25: 218–228.

Popper, K.R. 1963. *Conjectures and Refutations: The Growth of Scientific Knowledge.* New York: Routledge.

Powling, J.W. 2003. *The Mahogany Ship: A Survey of the Evidence.* Warrnambool: Osburne Group.

Prigogine, I. 1989. Thermodynamics and Cosmology. *International Journal of Theoretical Physics* 28: 927.

Prince, R. 2010. Policy Transfer as Policy Assemblage: Making Policy for the Creative Industries in New Zealand. *Environment and Planning A* 42 (1): 169–186.

Purcell, A.T. 1992. Abstract and Specific Physical Attributes and the Experience of Landscape. *Journal of Environmental Management* 34: 159–177.

Purcell, A.T. 1993. Relations Between Preference and Typicality in the Experience of Paintings. *Leonardo* 26 (3): 235–241.

Ranchordás, S. 2017. Digital Agoras: Democratic Legitimacy, Online Participation and the Case of Uber-Petitions. *The Theory and Practice of Legislation* 5 (1): 31–54.

Rawluk, A., and A. Curtis. 2016. Reconciling Contradictory Narratives of Landscape Change Using the Adaptive Cycle: A Case Study from Southeastern Australia. *Ecology and Society* 21 (1): 17.

Reason, P., and A. Bradbury. 2007. *Handbook of Action Research*, 2nd ed. London, UK: Sage.

Reason, P., and H. Bradbury (eds.). 2008. *Sage Handbook of Action Research: Participative Inquiry and Practice*, 2nd ed. London: Sage.

Redman, C.L. 2014. Should Sustainability and Resilience Be Combined or Remain Distinct Pursuits? *Ecology and Society* 19 (2): 37.

Redman, C.L., and A.P. Kinzig. 2003. Resilience of Past Landscapes: Resilience Theory, Society, and the *longue durée*. *Conservation Ecology* 7 (1): 14.

Reed, D., B. van Wesenbeeck, P.M.J. Herman, and E. Meselhe. 2018. Tidal Flat-Wetland Systems as Flood Defenses: Understanding Biogeomorphic Controls. *Estuarine, Coastal and Shelf Science* 213: 269–282.

Reise, K. 2002. Sediment Mediated Species Interactions in Coastal Waters. *Journal of Sea Research* 48 (2): 127–141.

Rockstrom, J., W. Steffen, K. Noone, A. Persson, F.S. Chapin III, E. Lambin, T.M. Lenton, M. Scheffer, C. Folke, H. Schellnhuber, B. Nykvist, C.A. De Wit, T. Hughes, S. van der Leeuw, H. Rodhe, S. Sorlin, P.K. Snyder, R. Costanza, U. Svedin, M. Falkenmark, L. Karlberg, R.W. Corell, V.J. Fabry, J. Hansen, B. Walker, D. Liverman, K. Richardson, P. Crutzen, and J. Foley. 2009. Planetary Boundaries: Exploring the Safe Operating Space for Humanity. *Ecology and Society* 14 (2): 32.

Saver, J.L., and A.R. Damasio. 1991. Preserved Access and Processing of Social Knowledge in a Patient with Acquired Sociopathy Due to Ventromedial Frontal Damage. *Neuropsychologica* 29 (12): 1241–1249.

Schirmer, J., M.L. Dare, and S.A. Ercan. 2016. Deliberative Democracy and the Tasmanian Forest Peace Process. *Australian Journal of Political Science* 51 (2): 288–307.

Scholz, R.W. 2011. *Environmental Literacy in Science and Society: From Knowledge to Decisions*. Cambridge, UK: Cambridge University Press.

Schön, D.A. 1991. *The Reflective Practitioner: How Professionals Think in Action*. London, UK: Basic Books Inc.

Schusler, T., D. Decker, and M. Pfeffer. 2003. Social Learning for Collaborative Natural Resource Management. *Society and Natural Resources* 15: 309–326.

Sherwood, J., I.J. McNiven, L. Laurenson, T. Richards, and J. Bowler. 2016. Prey Selection, Size, and Breakage Differences in Turbo Undulatus Found Within Pacific Gull (Larus pacificus) Middens Compared to Aboriginal Middens and Natural Beach Deposits, Southeast Australia. *Journal of Archaeological Science: Reports* 6: 14–23.

Sinclair, K., A. Curtis, E. Mendham, and M. Mitchell. 2014. Can Resilience Thinking Provide Useful Insights for Those Examining Efforts to Transform Contemporary Agriculture? *Agriculture and Human Values* 31: 371–384.

Sinclair, K., A. Rawluk, S. Kumar, and A. Curtis. 2017. Ways Forward for Resilience Thinking: Lessons from the Field for Those Exploring Social-Ecological Systems in Agriculture and Natural Resource Management. *Ecology and Society* 22 (4): 21.

Smith, A., and A. Stirling. 2010. The Politics of Social-Ecological Resilience and Sustainable Socio-technical Transitions. *Ecology and Society* 15 (1): 11.

Spies, T.A., E.M. White, J.D. Kline, A.P. Fischer, A. Ager, J. Bailey, J. Bolte, J. Koch, E. Platt, C.S. Olsen, D. Jacobs, B. Shindler, M.M. Steen-Adams, and R. Hammer. 2014. Examining Fire-Prone Forest Landscapes as Coupled Human and Natural Systems. *Ecology and Society* 19 (3): 9.

Steg, L., A.E. Van den Berg, and J.I.M. De Groot. 2013. *Environmental Psychology: An Introduction*. Chichester, UK: BPS Blackwell.

Stewart, J. 2006. Value Conflict and Policy Change. *Review of Policy Research* 23 (1): 183–195.

Strathern, M. 2004. The Whole Person and Its Artifacts. *Annual Review of Anthropology* 33: 1–19.

Trainor, S.F. 2006. Realms of Value: Conflicting Natural Resource Values and Incommensurability. *Environmental Values* 15: 3–29.

Turner, B.L., R.E. Kasperson, P.A. Matson, J.J. McCarthy, R.W. Corell, L. Christensen, N. Eckley, J.X. Kasperson, A. Luers, M.L. Martello, C. Polsky, A. Pulsipher, and A. Schiller. 2003. A Framework for Vulnerability Analysis in Sustainability Science. *Proceedings of the National Academy of Sciences* 100 (14): 8074–8079.

Turner, L.M., L.M. Turner, R. Bhatta, L. Eriander, L. Gipperth, K. Joannesson, A. Kadfak, I. Karunasagar, P. Knutsson, K. Laas, P. Moksnes, and A. Godhe. 2017. Transporting Ideas Between Marine and Social Sciences: Experiences from Interdisciplinary Research Programs. *Elementa: Science of the Anthropocene* 5 (14): 1–9.

Tweed, S.O., M. Leblanc, J.A. Webb, and M.W. Lubczynski. 2007. Remote Sensing and GIS for Mapping Groundwater Recharge and Discharge Areas in Salinity Prone Catchments, South-Eastern Australia. *Hydrogeology Journal* 15: 75–96.

Wagenaar, H. 2011. *Meaning in Action: Interpretation and Dialogue in Policy Analysis*. Armonk, NY and London, England: M.E. Sharpe.

Walker, B.H., N. Abel, J.M. Anderies, and P. Ryan. 2009. Resilience, Adaptability, and Transformability in the Goulburn-Broken Catchment, Australia. *Ecology and Society* 14 (1): 12.

Walker, B.H., S.R. Carpenter, J.M. Anderies, N. Abel, G.S. Cumming, M.A. Janssen, L. Lebel, J. Norberg, G.D. Peterson, and L. Pritchard. 2002. Resilience Management in Social-Ecological Systems: A Working Hypothesis for a Participatory Approach. *Conservation Ecology* 6 (1): 14.

Walker, B.H., S.R. Carpenter, J. Rockstrom, A.-S. Crépin, and G.D. Peterson. 2012. Drivers, "Slow" Variables, "Fast" Variables, Shocks, and Resilience. *Ecology and Society* 17 (3): 30.

Walker, B.H., L.H. Gunderson, A.P. Kinzig, C. Folke, S.R. Carpenter, and L. Shultz. 2006. A Handful of Heuristics and Some Propositions for Understanding Resilience in Social-Ecological Systems. *Ecology and Society* 11 (1): 13.

Walker, B., C.S. Holling, S.R. Carpenter, and A. Kinzig. 2004. Resilience, Adaptability and Transformability in Social–Ecological Systems. *Ecology and Society* 9 (2): 5–14.

Walker, B.H., and D. Salt. 2006. *Resilience Thinking: Sustaining Ecosystems and People in a Changing World*. Washington, DC, USA: Island Press.

Walker, B.H., and D. Salt. 2012. *Resilience Practice: Building Capacity to Absorb Disturbance and Maintain Function*. Washington, DC, USA: Island Press.

Watts, V. 2013. Indigenous Place-Thought and Agency Amongst Humans and Non-humans (First Woman and Sky Woman Go on a European Tour!). *DIES: Decolonization, Indigeneity, Education and Society* 2 (1): 20–34.

Wernli, D., and F. Darbellay. 2016. Interdisciplinarity and the 21st Century Research-Intensive University: Pushing the Frontiers of Innovative Research. LERU. https://www.leru.org/files/Interdisciplinarity-and-the-21st-Century-Research-Intensive-University-Full-paper.pdf. Accessed 17 Oct 2018.

West, S. 2015. Negotiating Social Ecological Fit Through Knowledge Practice. Licentiate Thesis in Natural Resource Management, Stockholm University.

West, S., R. Beilin, and H. Wagenaar. 2019. Introducing a Practice Perspective on Monitoring for Adaptive Management. *People and Nature* 1: 387–405.

White, D., A. Rudy, and B. Gareau. 2017. *Environments, Natures and Social Theory: Towards a Critical Hybridity*. London: Palgrave Macmillan.

Williams, K.J.H., M.A. Weston, S. Henry, and G.S. Maguire. 2009. Birds and Beaches, Dogs and Leashes: Dog Owners' Sense of Obligation to Leash Dogs on Beaches in Victoria, Australia. *Human Dimensions of Wildlife* 14 (2): 89–101.

Wilson, E. O. 1998. *Consilience: The Unity of Knowledge* (No. 31). New York: Random House Digital.